Science WORKBOOK 2

David Blaker, Anne Rowlands, Brent Smith

NELSON
A Cengage Company

Australia • Brazil • Japan • Korea • Mexico • Singapore • Spain • United Kingdom • United States

Science Workbook 2
1st Edition
David Blaker
Anne Rowlands
Brent Smith

Text designer: Cheryl Rowe
Cover designer: Cheryl Rowe
Production controller: Siew Han Ong
Reprint: Natalie Orr

Any URLs contained in this publication were checked for currency during the production process. Note, however, that the publisher cannot vouch for the ongoing currency of URLs.

Acknowledgements
Page 14, Maria Engelken; page 27, Nga Manu Images; page 33, Laurence Clark; page 42, Biosecurity New Zealand / MAF; page 57, Public Health Image Library; page 83, USAid in Africa; pages 132, 149, 150, 154 NASA Image Library; page 133, NIWA; page 139, GNS Science; page 148, Christopher J. Picking.

© 2011 Cengage Learning Australia Pty Limited

For product information and technology assistance,
in Australia call **1300 790 853**;
in New Zealand call **0800 449 725**

For permission to use material from this text or product, please email
aust.permissions@cengage.com

National Library of New Zealand Cataloguing-in-Publication Data
National Library of New Zealand Cataloguing-in-Publication Data

Blaker, David (David Neville)
Science workbook. 2 / David Blaker, Anne Rowlands, Brent Smith.

ISBN 978-017021-466-7
1. Science—Problems, exercises, etc.—Juvenile literature. [1. Science—Problems, exercises, etc.]
I. Rowlands, Anne. II. Smith, Brent. III. Title.
507.6—dc 22

Cengage Learning Australia
Level 7, 80 Dorcas Street
South Melbourne, Victoria Australia 3205

Cengage Learning New Zealand
Unit 4B Rosedale Office Park
331 Rosedale Road, Albany, North Shore 0632, NZ

For learning solutions, visit **cengage.com.au**

Printed in Singapore
10 11 12 13 25 24 23 22 21 20

The nature of science

What is science? It's our way of exploring and investigating the natural world, from atoms to animals, from earthquakes to energy. The process of discovery is not always straightforward. Science is always growing and changing, with a mixture of certain facts and uncertain information and ideas and theories. Why uncertain? In many situations – both living and non-living – we can't be sure because there is not enough evidence yet.

Science is built on three foundations: questions, ideas, and evidence. Without questions and curiosity there would be no progress. Without ideas and theories and imagination, science would just be a collection of facts. And as teams of scientists work at uncovering more evidence, ideas can become stronger.

Science is useful. It provides a platform for all technology and medical treatments, and also offers a basis for decisions on many social and environmental issues. It even helps you sort out sense from nonsense in aspects of everyday life.

Nature of Science is a description that sums up the main features that link the many parts of science. These features include questions, investigations, evidence, experiments, theoretical models, and practical uses. A *Nature of Science* strand runs through this book, and is a feature of many of the tasks.

The textbook SciencePLUS 2 contains detailed information, many practical experiments, more than 150 ideas for investigations and a student CD. Guidance is provided where units in this workbook link with units in SciencePLUS 2 (e.g. **SP2 Unit 3.8** = SciencePLUS 2 Unit 3.8).

Contents

Start-up

Life Science

Chemical Science

Physical Science

Earth Science

Astronomy

Glossary 156

Periodic table of the elements 160

For Students

This science workbook is intended for use either at home or in class. It can be used as a stand-alone resource, but is better used alongside the textbook *Scienceplus 2* – which has much wider information. Each activity in this workbook is linked to a particular unit in the textbook. You can do almost all workbook tasks at home without a textbook, but in most cases the task will be easier if you have recently covered the linked unit in class time. You will find that the activities range from easy to difficult. Some may take you less than 10 minutes, some more than 20 minutes. Most tasks need short answers. The amount of space provided for each answer gives an idea of how much to write. Harder tasks have the word **CHALLENGE** in the margin. A few of the challenge tasks are at the same standard as Level 1 science.

It also helps to be clear about the type of answer needed in each situation. It may help you to visualise that each question, activity and skill belongs in one or other level of a three-storey building. Mostly, you will be building knowledge on the ground floor – but the higher floors will become more important as your knowledge increases. How can you know which floor you are supposed to be working on? Look for key words, especially verbs like describe, explain, discuss.

Second floor: Judging and imagining
Look for any of these words:
Discuss, evaluate, imagine, judge, predict, design, plan, suggest, hypothesise.
EXCELLENCE

First floor: Understanding and applying
Look for any of these words:
Explain, compare, classify, sort, analyse, reason, summarise, arrange in sequence, apply a principle.
MERIT

Ground floor: Knowledge and remembering
Look for any of these words:
Describe, count, define, identify, list, match, name, observe, label, draw, select.
ACHIEVED

You will not find any research projects in this workbook, but *Scienceplus 2* has more than 190 ideas for investigations and projects, plus hundreds of review questions and 55 hands-on practical observations and experiments.

Start-up

If you need to revise information on safety rules and the names and uses of equipment, then this will be a good time to revise the Start-up section in *Science WORKBOOK 1*, and the introduction units in *Scienceplus 1*.

In this *Science WORKBOOK 2*, you could do all start-up activities in sequence at the beginning, or select some when the need arises. Example: You could postpone Start-up 2 and 3 until you need to plan experiments.

1 Questions

Date for completion: / /

Parent sig: _____
Teacher sig: _____

SP2 Unit 1.1

Questions, ideas, evidence: these are the foundations of science. Curiosity is part of human nature, and questions have been the starting point for almost all major science discoveries.

Next to each of the photos below and on the next page, make a list of five or more of your own questions based on that particular photo. To help get you started, two questions have been written next to the mosquito photograph.

Science

QUESTIONS   IDEAS   EVIDENCE

A

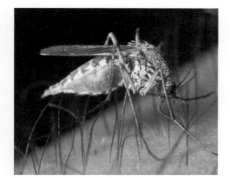

How do mosquitoes find you: by smell, or body heat?

What kinds of mosquitoes carry dengue fever?

_____

_____

_____

_____

B

_____

_____

_____

_____

_____

_____

ISBN: 9780170214667

C

_____
_____
_____
_____
_____
_____

D

_____
_____
_____
_____
_____
_____

| 2 | Science methods | | Date for completion: / / | Parent sig: _____ Teacher sig: _____ |

SP2 Unit 1.1/1.2

1 Complete the sentences below, using these 13 words: *same, controlled, proof, fair, independent, variable, repeated, evidence, theoretical, try-out, dependent, hypothesis, experiment.*

One way to sort out good science ideas from not-so-good ones is to put each idea in the

form of a _____ (1), which is a statement that can be tested. The

_____ (2) might support the hypothesis, or it might not. Lots of good quality

evidence can add up to _____ (3). One way – but not the only way – to get

good evidence is by _____ (4). In a scientific experiment, which really just means

a _____ (5) , we generally aim to change just one _____ (6) at

a time. The _____ (7) variable is the one that you choose to change during

the experiments. The _____ (8) variable is the one that is recorded as part of

your results. All other variables should if possible be kept the _____ (9), so

that comparisons are _____ (10). Variables that are kept constant throughout

the experiment are called _____ (11) variables. Each trial should, if possible,

be _____ (12) several times so we can be sure of the result. If repeated results

don't fit in with the _____ (13) model, both the results and the model need to

be checked.

2   The word 'model' has three quite different meanings:
    - a **good example**
    - a **scale model**: which can be bigger or smaller than the original
    - a **theoretical model**: an overall picture of how things work.

Classify each of the following examples of a model into one of the above meanings:

A   Have a look at this model of a plant cell. _____

B   Continental drift models revolutionised understanding. _____

C   She could be a fashion model if she wanted. _____

D   He's a role model for the rest of the team. _____

E   Atomic theory models have changed in the past 100 years._____

F   This is a model of the solar system. _____

G   Climate change models predict that the Amazon rainforests could begin to dry out.

    _____

3   A fair trial?

Date for completion: / /   Parent sig: _____   Teacher sig: _____

SP2 Unit 1.2/1.3

Dan keeps pet rats. He has noticed that sometimes they build big nests, at other times small nests. He wonders what's behind this, then tries out his hypothesis that nest-building behaviour depends on surrounding temperature. He puts one of his tame rats in its own safe cage, gives it a pile of straw in the opposite corner of the cage to its empty nest box, and keeps the rat in its cage outside in the carport for a week. At night, outside air temperatures fall as low as 5 °C. At the end of the week he takes out all the rat's nest material and weighs it to the nearest gram. He then starts again with the same rat and a new lot of straw, and for the next week keeps the rat cage in his bedroom, where the temperature averages about 20 °C. At the end he weighs the nest again.

1   From the above description and A to J below, draw what you think is the cage layout, and label the different features.

ISBN: 9780170214667

2   From the above description, draw a 2-week timeline and use it to mark in the main events of Dan's rat experiment.

3   Decide whether each of the following is a **dependent variable**, an **independent variable**, a **controlled variable**, or **irrelevant** to his experiment.

A   The kind of cage he kept the rat in._____

B   The amount of nest material provided. _____

C   The type of nest material provided. _____

D   The surrounding air temperature. _____

E   The rat's age. _____

F   The rat's name. _____

G   The type of food he gave it. _____

H   The weight of nest material at the end. _____

I   Whether the rat is male or female. _____

J   The rat's water supply. _____

4   Suggest one possible improvement to Dan's rat experiment.

_____

_____

---

4   Big words

Date for completion: / /   Parent sig: _____
Teacher sig: _____

SP2 Unit 1.4

In science, especially in medical science, some words are based on classical Latin and Greek languages. Science is not about learning lots of big words, but if technical words are a headache for you, it may help to understand how they are put together. Example: the word 'hydrogen' is from *hydro* (water) + *gen* (to make), because when you burn hydrogen you get water.

Here is a list of technical words. Explain the meaning of each, based on the Greek and Latin meanings given in the box on the next page.

A   What is *palaeontology*? _____

B   What does *cardiologist* mean? _____

C   What does a *haematologist* specialise in? _____

D   A *dermatologist* specialises in_____

E *Osteosarcoma* means cancer of the_____

F *Geology* literally means _____

G *Hydrology* is a branch of science that specialises in_____

H A *seismometer* measures what? _____

I What does *archaeology* mean? _____

J What does a *microbiologist* specialise in? _____

K What does a *petrologist* specialise in? _____

L What does *laryngitis* affect? _____

M *Rhinoplasty* changes the shape of what? _____

N An *otorhinolaryngologist* specialises in: _____

O What colour is the mineral *haematite*? _____

P Where does a *hypodermic* injection go? _____

Q What does *hypothesis* literally mean? _____

---

archae: ancient	*haem*: blood	*mikros*: small	*petra*: rock or stone
cardio: heart	*hydro*: water	*os*: bone(s)	*rhino*: the nose
derm: skin	*hypo*: under	*otos*: the ear	*seismos*: earthquake
geo or *ge*: Earth	*larynx*: throat/voicebox	*palaios*: ancient	*thesis*: a suggestion

Words with '-ology' come from *logos*, the Greek word for discussion/understanding.

---

5 Fact or what?

Date for completion: / /   Parent sig: _____
Teacher sig: _____

SP2 Unit 1.4

Science has a mixture of observation, evidence, inference and prediction. What's the difference?

Observation: what you can see, hear, feel, measure.
Evidence: facts or signs showing that something exists or is true.
Inference: something you think is true, based on evidence.
Prediction: a statement about what you think is going to happen in the future. Some predictions turn out to be correct, some not. It helps to have good reasons!

The four boxes below each contain one observation, one prediction, and one inference (O, P, I). First, fill in the column on the right. Next, put the matching left-side letters in the spaces underneath, to find what Leo says about his family pet.

C	Look at those clouds building up.	O, P, or I?
H	It's probably going to rain this afternoon.	
L	The humidity must be high.	

E	You could get hurt if he gets off the see-saw now.	
A	You're high up on the see-saw and he's at ground level.	
I	Because he's heavier than you are.	

ISBN: 9780170214667

T	The test tube feels warm.	
K	The reaction must be generating heat.	
A	If we add more magnesium it should get even hotter.	

E	Everything is the teacher's fault.	
S	The test results are not good.	
T	Results will improve when we get a new teacher.	

Observations: Inferences: Predictions:

— — — —   — — — —   — — — — , says Leo

Flexible or fixed?

Science is not just a collection of facts; it is more a process of discovery. It is about asking questions, coming up with ideas, testing predictions. Also, not everything in science is definite. Ideas can change over time; some are almost completely certain, some less certain. Empirical scientific methods help find which ideas are backed by good evidence, and which are not.

Empirical (adjective): any knowledge based on observation, experience, evidence, and testing; not based only on opinions and arguments.

6 Certain or not?

Date for completion: / /
Parent sig: _____
Teacher sig: _____

SP2 Unit 1.1–1.4

Much of science is based on proven facts. If it wasn't, then aircraft would not fly and medicines would not work. But not everything in science is absolutely certain. In many areas of knowledge there is uncertainty, with more questions than answers.

- What is an atomic nucleus made of? Nobody is sure. There's no final answer yet.
- How do memories get stored in your brain? We're not sure. They seem to move from place to place.

Many facts are 100% certain, like $E = mc^2$, but many areas of knowledge are still being sorted out. As new evidence comes in, some science theories become strengthened, and other theories are changed. Example: 70 years ago 'continental drift' was a strange new idea. Because of growing evidence, the idea is now 100% certain, although many details may never be known.

On the next page is a list of scientific-sounding statements and predictions. Many are almost 100% certain, some less certain, some very uncertain. Place each letter A to T on the right, where you think it best belongs on the 100% to 0% line. Even though you don't have all the information, be able to justify your choices.

A all oxygen in the air has been put there by
green plants

B Earth could easily support and feed 12 billion
people

C predicting where major earthquakes are
likely to happen

D predicting when major earthquakes are likely
to happen

E smallpox has been completely eliminated
from the whole world

F genetic engineering of crop plants is a risk-
free technology

G $E = mc^2$

H electric cars will completely replace petrol-
driven cars by 2050

I the chance of next summer being drier than
average

J humans first began in Africa, and spread
from there

K smoking is a major cause of poor health

L reptiles appeared on Earth over 100 million
years before mammals did

M the amount of CO_2 in the air is rising each
year

N the chance of finding a human fossil in
Jurassic-age rocks

O the amount of crude oil underground is
running out rapidly

P science will find a substitute that will replace
oil as a source of energy

Q rocks at the top of Mount Everest were
originally formed under water

R our understanding of protons and neutrons is
not likely to change

S Pacific people originated somewhere in
South East Asia

T New Zealand has been occupied by people
for less than 1,000 years

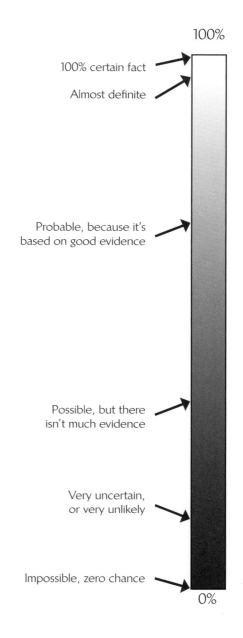

100%

100% certain fact

Almost definite

Probable, because it's
based on good evidence

Possible, but there
isn't much evidence

Very uncertain,
or very unlikely

Impossible, zero chance

0%

7 Quality of evidence

Date for
completion: / /

Parent sig: _____
Teacher sig: _____

SP2 Unit 1.4

The three foundations of science are questions, ideas and evidence. Without a basis of evidence, ideas
become wobbly. Whether in astronomy or economics or health, some evidence is strong, some is
medium, some is weak.

- **STRONG**. The subject has been investigated many times. Almost all results agree.
- **Medium**. One or two reports from reliable researchers.
- Weak. Rumours. One-example reports. Anecdote. Bias. Opinion stated as fact.

ISBN: 9780170214667

When doing a research assignment, you have to decide which facts and which evidence to use, and which to throw out. It helps to be aware that quality of evidence is not all the same. For each of the examples below, assess the quality of the evidence. Write its classification as: *strong, medium,* or *weak.* You might decide to classify some examples as in-between, such as *medium / strong.*

1   My uncle smoked all his life and lived to be 70 years old: that proves smoking does no harm.

_____

2   Dozens of studies show that eating omega-3 oils in food is linked to improved mental performance.

_____

3   We think smallpox has been eliminated, because no case has been seen worldwide since 1976.

_____

4   A coal industry spokesman says that coal is not contributing to the climate change problem.

_____

5   Some people have found that if they eat organic foods there is a general improvement in their health.

_____

6   People say that computers are making everybody more clever.

_____

7   A recent magazine article says that nuclear energy is completely safe if the power stations are maintained properly.

_____

8   Five independent studies have found that children who are given fizzy drinks from an early age are more likely to develop diabetes.

_____

9   Experience shows that brick buildings are more at risk in earthquakes, compared to buildings made of wood.

_____

10 Nobody in our family has been immunised against meningitis, and nobody has become sick with it. That proves immunisation is a waste of time.

_____

Life Science

1 No two alike

Date for completion: / /

Parent sig: _____

Teacher sig: _____

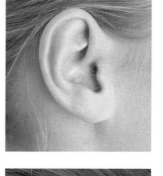

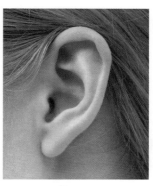

1 Each of us has many characteristics (features) that are mainly or totally controlled by our genes.

- Some inherited features show **either/or variation**. There are few in-betweens.
- Other features show **continuous variation**: over a large number of people there is a full range of in-betweens.

Decide which category (either/or?; continuous?) each of the following characteristics falls into.

Height _____

Earlobes attached or not _____

Hair colour _____

The ability to roll your tongue _____

Shoe size _____

Skin colour _____

Inheriting type-1 diabetes _____

General body shape _____

Hitchhiker's thumb _____

Human intelligence _____

Height: An example of continuous variation

Hitchhiker's thumb

Number of adult males (y-axis)

Height in metres (x-axis): 1.4   1.6   1.8   2.0

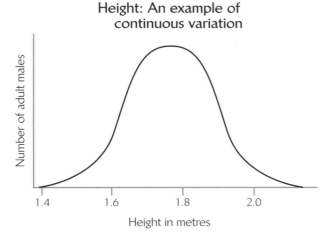

Tongue roll

ISBN: 9780170214667

2 With features that show continuous variation, some are influenced purely by **genes**, some partly by **environment**. 'Environment' means physical surroundings, but can also include food, education, social upbringing. Give each of the following characteristics a total of 4 ticks, e.g. 4 for genes and 0 for environment, or 1 and 3, or 2 and 2. (Your answers will be a matter of opinion.)

Characteristic (feature)	Total score out of 4	
	Genes	Environment
Your height		
Your ability in maths		
Your accent		
How fast you can run		
Freckled skin or not		
Hair colour		
Singing ability		
Your overall health		
A dog's colour		
How friendly a dog is to strangers		
Milk production in cows		
The overall shape of a pine tree		

2 Code breaking

Date for completion: / /

Parent sig: _____
Teacher sig: _____

SP2 Unit 2.2

Chemically, DNA molecules are long, thin and twisted into a double helix shape. One function of DNA is to carry coded information from one generation to the next. There are many kinds of other coded information in our lives, from mathematics to music.

Below are three examples of coded messages. Translate each message into English words.

1 Here is a text message. Translate it. Write the words on the line below.

7777887772255599908443377733068877778022330207788442225533777092999111

_____

2 Each letter in this message has been displaced by two characters from the original version. Translate the message, and write the original correct version on the line underneath.

NQICP   YCUJGU   KP   RQTTKFIG

_____

3 Work out the basis for this code. Translate the message.

4,15,21,2,12,5   8,5,12,9,24

_____

_____

_____

DNA

Date for completion: / /

Parent sig: _____

Teacher sig: _____

SP2 Unit 2.2/2.3

Life Science

1 The diagram below shows a short section of DNA. Complete the labelling by writing in the six empty boxes.

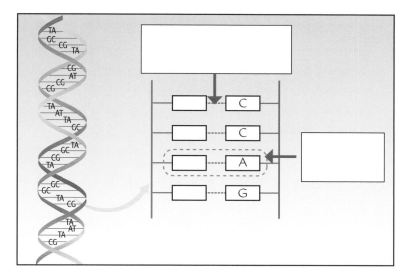

2 One key feature of DNA is its A-T and C-G pairing. This pairing exists because of the shape of these molecules. Complete the full names of these four molecules:

A _____   T _____

C _____   G _____

3 The drawing below shows DNA being replicated (i.e. doubled, or copied) just before cell division. Complete the drawing by writing in the letters A, T, C and G in their correct places.

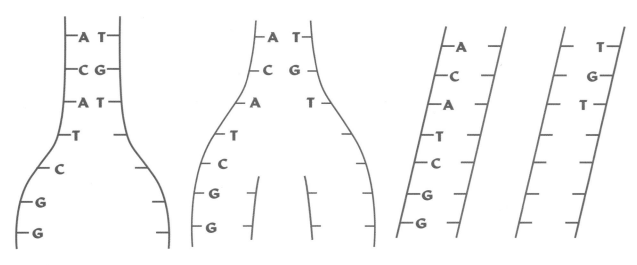

ISBN: 9780170214667

Ordinary cell division, also known as **mitosis**, makes two cells each containing DNA identical to the original cell's DNA. These four drawings **A**, **B**, **C**, **D** show different stages of mitosis in a cell that had two pairs of chromosomes at the start. Carefully draw in the chromosomes for stages **B**, **C** and **D**, keeping the same colour coding. **B** is already part-completed.

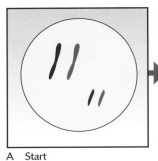

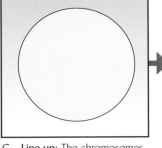

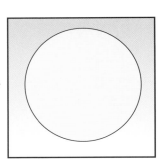

A   Start

B   Double up: The cell makes two identical copies of all its DNA. The chromosomes are doubled.

C   Line up: The chromosomes are arranged along the cell 'equator'.

D   Split up: An identical set of chromosomes moves towards each end of the cell.

5 **Mitosis and meiosis**

1   Write in the missing words. Choose from this list: *zygote, haploid, gametes, variety, diploid, chromosome*.

Meiosis is a special form of cell division that (in animals) occurs only in sex organs. Meiosis

produces eggs and sperm, also known as _____ (1). Meiosis

reduces the chromosome number so that eggs and sperm always have half the full number

of chromosomes, the _____ (2) number, **n**. This is so the fertilised

egg, the _____(3), will have the full number of chromosomes:

also known as the _____ (4) number, **2n**. As well as halving the

_____ (5) number, meiosis creates genetic _____ (6),

which causes no two sex cells to have exactly the same DNA.

2   Complete this table, choosing your answers from: *1, 2, 4, 23, 46, yes, no, millions, sex organs, growing tissues*.

	Mitosis	Meiosis
Happens where?		
Number of cells made from one cell?		
Number of chromosomes in each cell after cell division (in humans)?		
Are the final cells identical or not?		

Some terms in genetics are based on Greek words:
gamos = marriage;
di = two;
chroma = colour;
soma = body;
meiosis = lessening;
zygos = egg yolk.

Genetics crossword

SP2 Unit 2.1–2.16

Life Science

CHALLENGE

1   Choose any 9 to 12 words from this word list and use them to create your own crossword in the grid below. Outline the squares for your words and give each word a number. Underneath, write your own numbered clues for your chosen words.

variation	replication	fertilisation	mitosis
environment	meiosis	diploid	dominant
codon	recessive	homozygous	heterozygous
genome	alleles	genotype	phenotype
chromosome	mutation	bodycell	gamete
zygote	crossingover	doublehelix	gene

2   Exchange a blank crossword and your written clues with another student. Test each other.

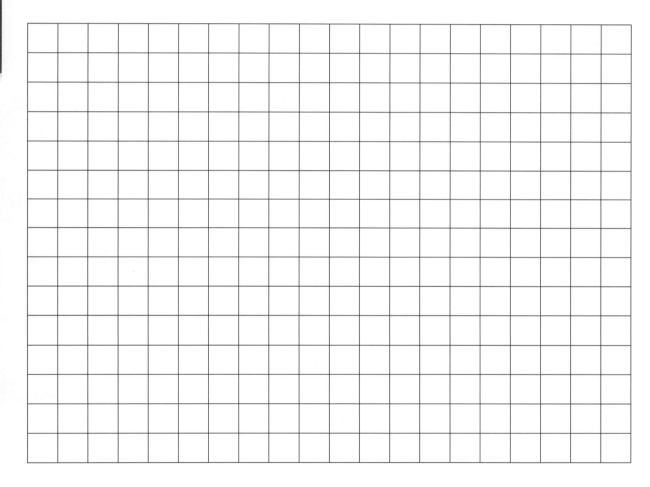

ISBN: 9780170214667

SP2 Unit 2.5

Date for completion: / /   Parent sig: _____
   Teacher sig: _____

Life Science

1   Draw lines to match each word on the left to the symbol or meaning that best describes it. One has been done for you. You may use each meaning only once.

Word
Heterozygous
A dominant allele
Homozygous recessive
Allele
Genotype
Homozygous dominant
Phenotype

Meaning
rr
The gene make-up of an individual
The outward result of genotype
RR
One version of a gene
R
Rr

2   Humans have a pair of genes for **hitchhiker's thumb** (the name for a 'curvy' thumb). Brock has one dominant allele (represented by the letter *H*) for hitchhiker's thumb and one recessive allele (represented by the letter *h*) for **non**-hitchhiker's thumb. This makes his genotype *Hh* for hitchhiker's thumb. What is his thumb phenotype?

Phenotype:_____

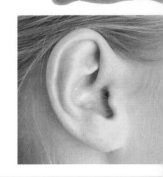

3   Having **unattached earlobes** is a dominant trait (gene-controlled feature) in humans. Christine has **attached** earlobes. What must her genotype be for this trait? The alleles are represented by **E** and **e**.

Genotype: _____

---

8 **Labrador colours**

SP2 Unit 2.5

Date for completion: / /   Parent sig: _____
   Teacher sig: _____

In labrador retrievers, one gene with two alleles decides coat colour. The yellow-coat allele (*D*) is dominant over the brown-coat allele (*d*).

1   A homozygous yellow-coat female (*DD* genotype) mates with a heterozygous yellow-coat male (*Dd*). Work out the different possible genes and colours of their offspring (pups) by completing the Punnett square on the next page. Start by writing the genotype inside each gamete (egg and sperm). Remember that gametes have only one or other form of each gene, never both. Two eggs are drawn in, but in this case both have the same genes (alleles).

	Egg ⬇	Egg ⬇
Sperm	Fertilisation result:	Fertilisation result:
Sperm	Fertilisation result:	Fertilisation result:

What are the pup *genotype* possibilities? _____

_____

What are the chances of this yellow-coat female and yellow-coat male producing brown-coat pups?

_____

2 In another mating between another pair of individuals, two heterozygous yellow-coat labrador parents produce eight pups. Work out the likely colours of their pups, using the Punnett square below.

Genotype of the father _____

Genotype of the mother _____

State the likely colours of their pups, and the likely numbers of each colour:

_____

Family genes

Date for completion: / / Parent sig: _____ Teacher sig: _____

SP2 Unit 2.5

The ability to roll your tongue into a tube is a purely inherited feature, governed (decided) by one gene with two alleles. Either you can roll your tongue, or you can't. On average, about one person in three is a non-roller. The allele for tongue-rolling (represented by the letter *T*) is dominant over the allele for non-rolling (represented by the letter *t*).

In the family tree diagram on the next page, females are represented by circles, males by squares. Each mother and father is directly linked by horizontal lines, and their children are shown by vertical links. In this diagram, rollers are shown with a **U**, non-rollers with a wavy line ~

ISBN: 9780170214667

James   Susan
U   U

KEY
U = Tongue-roller
~ = Non-roller

Ryan   Abby   Tessa   John   Daniel   Kiri
U   ~   ~   ~   U   ~

Zara   Jack   ?   ?   Katie   Jake   Sam   Nicky
~   ~   U   U   U   U

Using the above information, work out what must be the genotype of everyone in the family. Each person's gene arrangement is *TT* or *Tt* or *tt*.

1   How many children do James and Susan have? _____

2   Who does Kiri marry? _____

3   What are the genotypes of Abby, Tessa, John, Zara, Jack and Kiri? _____.
    Give one reason for deciding that all these individuals have this genotype.

    _____

    _____

    _____

4   What are the tongue-roll genotypes of James and Susan? _____
    Explain your reasons for saying this.

    _____

    _____

5   What is the phenotype of the twins (for tongue-roll)? _____
    Explain your reasons for saying this.

    _____

    _____

    _____

6   What is the genotype of Katie? _____
    Explain your reasons for saying this.

    _____

    _____

7   Write each individual's genotype inside their circle or square. Where it is not possible to be 100% certain, put the most probable genotype in brackets.

8   Daniel's possible tongue-roll genotypes are _____ and _____.

    Which do you think is more likely?_____ Explain your reasons.

    _____

    _____

10 Genetics puzzle

Date for completion: / /

Parent sig: _____

Teacher sig: _____

SP2 Unit 2.1–2.16

1 Write the missing words in the spaces below. Select from this list: *mutation; DNA; mitosis; phenotype; segments; genes; recessive; environment; diploid; continuous varitation; nucleus; sex; heterozygous; chemicals; inherited; 23; meiosis; chromosomes; traits; half; radiation; written; parent.*

Variation is caused by a combination of the _ _ _ _ _ we inherit and our _ _ _ _ _ _ _ _ _ _ _.
                                           46 31   2   26

Features like height, weight and skin colour show _ _ _ _ _ _ _ _ _ _ _ _ _ _ _ _ _ _.
                                                   5   36   28   33 48

Inside a cell you will find the control centre, or the _ _ _ _ _ _ _. Inside this we can
                                                        11 34

find _ _ _ _ _ _ _ _ _ _. Each one of these is made up of _ _ _ _ _ _ _ _ called
     1 29 7 40   10   16

genes, which in turn are made of coiled up _ _ _ .
                                            4

_ _ _ _ _ _ are physical characteristics that can be passed down or _ _ _ _ _ _ _ _ _ from
9 21   50   42 23   47

one generation to the next.

When regular body cells divide, we call it _ _ _ _ _ _ _.
                                           30 44   43

The full number of chromosomes in a cell (in humans it is 46) is called the _ _ _ _ _ _ _ number. This
                                                                             39   35 12

can be _ _ _ _ _ _ _ as 2n.
       3/41

The cell division that occurs in _ _ _ cells is called _ _ _ _ _ _ _. In this process you
                                 18   15   45

reduce the number of chromosomes to _ _ _ _ (in humans that would be _ _ chromosomes).
                                     14   20

Most organisms inherit genes in pairs, one from each _ _ _ _ _ _.
                                                      24

A pair of genes consisting of one dominant and one recessive allele is said to be

_ _ _ _ _ _ _ _ _ _ _ _.
13   25   32

A _ _ _ _ _ _ _ _ _ is the outward appearance of a gene, like blue eyes.
        6

A gene which is not expressed if a dominant gene is present, is said to be _ _ _ _ _ _ _ _ _.
                                                                           17

A _ _ _ _ _ _ _ _ is a mistake in the DNA sequence. They can be genetic, or caused by
  8   27

environmental factors such as _ _ _ _ _ _ _ _ _ and _ _ _ _ _ _ _ _ _ such as nicotine.
                              19   22 49   37/
                                                   38

2 Use the numbered letters from your answers above to solve the riddle below.

_ _ _   _ _   _ _ _   _ _ _ _ _ _   _ _ _ _   _ _ _ _   _ _
1 2 3   4 5   6 7 8   9 10 11 12   13 14 15   16 17 18   19 20 21

_ _ _ _ _ _ _ _ _ _ ?
22 23 24 25 26 27 28 29 30 31

_ _ _   _ _ _ _ _   _ _ _ _   _ _ _ _ _   _ _   _ _ _ _
32 33 34   35 36 37 38   39 40 41 42   43 44 45   46 47   48 49 50!

ISBN: 9780170214667

11 Mutations

SP2 Unit 2.6

Date for completion: / /
Parent sig: _____
Teacher sig: _____

1 Complete these sentences. The missing words are provided in this list: *variation, radiation, mutant, base, DNA, chromosome, gamete, white, webbed, body.*

A mutation is like a spelling mistake that alters messages carried in genes. A mutation

may be a small as changing one single _____ (1) in one section of

_____ (2), or as big as adding a whole extra _____ (3).

A _____ (4) is the word for any plant or animal that has had a mutation. Albino

mutants exist in many different animals, resulting in their skin colour being almost completely

_____ (5). A mutation can only be passed on to the next generation if it

affects sex cells; mutations that affect only _____ (6) cells cannot be passed

on. Many mutations are harmful, but some are useful, such as _____ (7) feet

in swimming animals. Mutations are an important part of evolution, because they increase the

amount of genetic _____ (8). Mutations can be caused by some kinds of

_____ (9) and also by some kinds of 'mutagen' chemicals. A mutation that

occurs in a _____ (10) is likely to be passed on to children.

12 More mutations

SP2 Unit 2.6

Date for completion: / /
Parent sig: _____
Teacher sig: _____

1 Which of these six features could possibly be inherited, even partly, in the DNA of a person's children? Answer yes or no in each case.

having pierced ears _____

having a dark tan from much time in the sun _____

getting melanoma cancer from too much time in the sun _____

speaking the language they grew up with _____

having green eyes _____

being blind as a result of injury _____

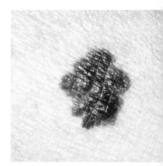

A small melanoma.

2 A section of DNA normally carries the message

   AAGTCTAGCGTATTCGCG

Below are four different mutations that change the message. Underline the affected bases in each case. In each case also classify the mutation as either: **deletion**, **insertion**, or **repetition**.

   (1)   AAGTCTAGCGTATTCGGCG _____

(2)   AAGTCTAGCGTATATATTCGCG _____

(3)   AAGTCTAGCGTATTCGG _____

(4)   AAGGTCTAGCGTATTCGCG _____

Date for completion: / /

Parent sig: _____

Teacher sig: _____

SP2 Unit 2.7/2.8

Life Science

1   The word 'theory' has two different kinds of meaning:

· an **opinion** or a guess (A)

· a **big idea** that explains many facts. (B)

Decide in which sense the word 'theory' is used in each of the following statements. Write A or B in each space provided.

In theory they should arrive tomorrow after breakfast _____

In music theory the main elements are melody and rhythm _____

In theory they should have won the game easily _____

Evolutionary theory explains many aspects of life on Earth _____

It's not hard to understand the basics of atomic theory _____

A great amount of evidence points to the conclusion that all mammals share the same ancestors. This diagram shows one kind of evidence: the front limb skeletons of three mammals.

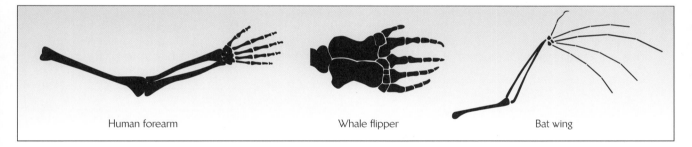

Human forearm   Whale flipper   Bat wing

2   Identify three ways in which all three drawings show the same bone arrangements.

_____

_____

_____

_____

_____

3   Describe one major difference between the bone proportions of bat wing and whale flipper.

_____

_____

_____

_____

ISBN: 9780170214667

Date for completion: / /

Parent sig: _____

Teacher sig: _____

SP2 Unit 2.7/2.8

Life Science

The idea of evolution is an old one, but Charles Darwin was the first to suggest a theory of how changes and adaptations could happen in nature. Supply the missing words in the summary below by choosing from this list: *environment; individuals; variation; survival; selection; adaptations; reproduce; offspring; genes; identical; young; genetics; successfully.* Some words may be used more than once.

In the wild, most animals die _____(1). Most, like the fish in the photo,

starve or are killed before they get a chance to _____(2). It is a fact that

animals produce more offspring than the _____(3) can support. It is also a

fact that in any population of animals there is a lot of genetic _____ (4),

and it is rare for any two individuals to be _____(5). On average,

_____ (6) that are better fitted to their _____ (7)

are likely to have more _____ (8). As a result, some

_____ (9) become commoner than others, over time. This process is

known as natural _____ (10). Although Charles Darwin did not know

much about _____ (11), his basic ideas have since been confirmed

and evolutionary theory has grown and become accepted as the best way of explaining

how living things change. Natural _____ (12) is sometimes known as

_____ (13) of the fittest, but the word 'fit' can be misleading. 'Fit'

does not only refer to animals that are big and strong, 'fit' refers to those that because of their

_____ (14) are able to reproduce more _____(15).

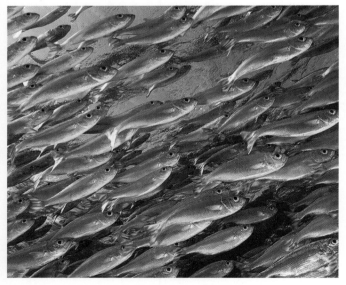

These adult fish are the survivors of a much greater number of young fish. They may all look the same, but they are not identical.

'Why ?'

Often we hear the question *why*, especially about animal adaptations. Examples: Why are these fish that colour? Why do they migrate?

It helps to realise that 'why' can be three separate questions.

1 What is the **cause**? (What makes it happen? Weather, seasons, hormones?)
2 What is the **function**? (What use is it? How does it help the animal live longer?)
3 How could it have **evolved**?

Function and evolution questions only apply to living things, not chemicals.

ISBN: 9780170214667

Kakariki

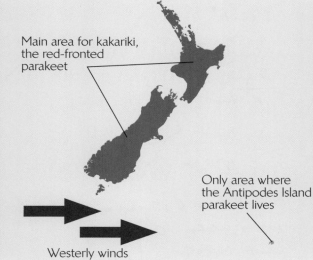
Main area for kakariki, the red-fronted parakeet

Only area where the Antipodes Island parakeet lives

Westerly winds

Antipodes Island parakeet

Life Science

Several kinds of kakariki (parakeet) are native to New Zealand. The red-fronted species occurs throughout the country, including some offshore islands. The remote Antipodes Island is home to a similar but different species: larger, coloured differently, and a weak flyer. The following events A to F could explain how both species came to be now living on Antipodes Island. The events are not in the correct order.

A Over time, the isolated Antipodes birds became genetically very different to the South Island birds.

B Once upon a time kakariki occurred in the South Island, but none existed on Antipodes Island.

C Natural selection caused shorter-winged, bigger individuals to become more common on Antipodes Island.

D Land birds sometimes get blown out to sea during storms. Thousands of years ago, some wind-blown parakeets arrived on Antipodes Island.

E More recently, some red-fronted kakariki arrived from the South Island. The Antipodes Island parakeets were by now so different that the two kinds could not interbreed.

F In their stormy new environment, birds that flew more actively were more likely to be blown away from Antipodes Island, and died at sea.

1 Read the events carefully, and list them here in the order in which they probably took place.

First ☐ ☐ ☐ ☐ ☐ ☐ Most recent event

2 Add arrows and words to the map above, to summarise the main events in fewer words than the A to F summary.

ISBN: 9780170214667

Date for completion: / /

Parent sig: _____

Teacher sig: _____

Life Science

This is a page from Charles Darwin's notebook, made in the Galapagos Islands when he first began to think about a theory of evolution. It shows a branching arrangement. Some older books show evolution looking like a ladder, with animals at different stages and moving upwards to 'higher' stages. All the evidence suggests that evolution is like a tree, not a ladder.

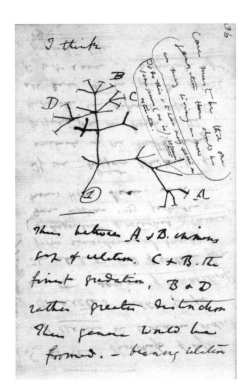

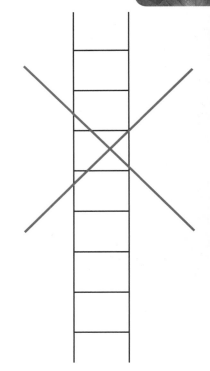

Ten different groups of mammals:

sheep   deer   cats   mice   rats   whales   guinea pigs dolphins kangaroos possums.

Draw a tree-shaped diagram showing how these 10 different mammal groups might be linked, with all having the same ancestors at the start. Animals that are physically similar should be placed on nearby branches with very different animals on different branches. (Your answer may be based on insufficient facts but that is OK. Science depends on ideas as well as on facts.)

?

↑

Probable mammal ancestors

SP2 Unit 2.10

Reducing agricultural greenhouse gas emissions is an important goal – and scientists are on a mission to design inhibitors that will shut down methanogens – bacteria that produce methane inside the rumen (stomach) of sheep and cows. The methane is then burped out – and is a much more potent greenhouse gas than CO_2 is.

AgResearch scientists Dr Attwood and Dr Ronimus are looking to reduce methane production by designing molecules that inhibit enzyme reactions essential to methanogen function.

These two scientists and their teams firstly needed to identify features shared by different methanogens. Their successful sequencing of the methanogen genome has allowed them to identify some shared DNA features.

The trick is to design inhibitors that only knock out the methanogens – not other

organisms in the rumen which digest food and supply nutrients for the animal. The scientists aim to find genes that are shared by methanogens, but are not closely related to other organisms.

They go about this by comparing genetic sequences and examining metabolic pathways to identify genes found in methanogens only. Once they've identified those genes, then they look at the proteins coded by these genes.

Drs Attwood and Ronimus are looking for the catalytic site - the engine room of the enzyme where the reaction actually takes place. Having identified this site and how it is structured, they can then go about designing a compound to inhibit this reaction.

Enzymes can be described as a lock with a key, where the enzyme is the lock and the key the substrate that fits into that lock. The hunt is on to find another key that will fit the lock, and prevent the enzyme from working.

It is a complex design process, but they have made progress towards cracking it. They are optimistic about building methane-inhibiting compounds that can be delivered to sheep and cows, and so help reduce agricultural greenhouse gas emissions.

(Information from agresearch.co.nz)

Life Science

For each of the eight paragraphs above, write two short bullet points that summarise the information.

_____

_____

_____

_____

_____

_____

_____

_____

_____

_____

_____

ISBN: 9780170214667

18 Ancestors

Date for completion: / /

Parent sig: _____

Teacher sig: _____

Life Science

This kind of drawing is known as a **pedigree**. It shows the direct ancestors (tipuna) of any one individual, but does not show brothers and sisters. Each of us inherits genes from many ancestors. Nobody inherits all or even most of their genes from just one or two ancestors.

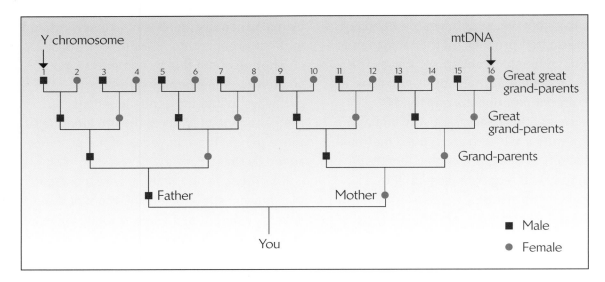

1 According to the diagram above, about what fraction of your genes are inherited from each great-great-grandparent? _____

2 How many great-great-grandmothers do you have? _____

3 The Y-chromosome is one of the 46 we have, and occurs in males only. According to the diagram, from which ancestor did 'You' inherit his Y-chromosome?

_____

4 mtDNA (mitochondria DNA) is inherited in a different way. From which one person in this pedigree did 'You' inherit all your mtDNA? _____

5 Counting your parents as one generation back, state how many ancestors you are descended from:

two generation back _____, three generations back _____,

four generations back _____, five generations back _____, six generations

back _____, 7 generations back _____.

CHALLENGE

6 If you continue the calculation for 27 generations (about 800 – 1000 years), you find that theoretically you are decended from 536,870,912 ancestors. That is more people than were alive in the whole world back then. How can you explain the 'extra ancestors'? (No, it's not aliens.)

_____

_____

_____

Date for completion: / /

Parent sig: _____
Teacher sig: _____

SP2 Unit 2.11

Life Science

Until recently, it was believed that humans were in separate races: black African, white European, Polynesian, Asian, etc; with little connection between the races. With new DNA evidence, the picture has changed.

All males have one Y-chromosome. This DNA is only inherited from father to son. Other DNA, known as mtDNA, is passed only down the female side of the family. Analysing both kinds of DNA and the rate they change has led to an amazing discovery: all human beings are descended from one small group – possibly even from one single female, though that is impossible to prove scientifically. These people lived somewhere in East Africa – probably around 100,000 years ago.

Perhaps around 70,000 years ago humans first began to leave Africa. There is only one human race, and we all originated in Africa. All humans are related to each other, most of us sharing ancestors from as recently as a few thousand years back.

1 Underline the main points in the article above.
2 Write a summary in 8 to 12 bullet points.

_____
_____
_____
_____
_____
_____
_____

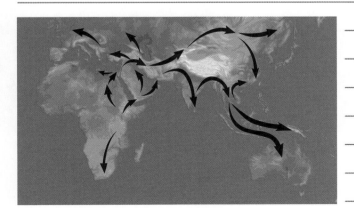

_____
_____
_____
_____
_____
_____
_____
_____

ISBN: 9780170214667

Life Science

Compared to the rest of the world, the Pacific islands were populated quite recently, and it happened very fast. Evidence from DNA points to two separate waves of migration from Asia. People first arrived in Australia and New Guinea about 40,000 years ago.

Another surge of migration began about 4,500 years ago. This involved small groups of migrants originating from what we now call Taiwan. Evidence? Some DNA markers are common in Pacific people and some tribes in Taiwan, but not common anywhere else in the world. There are also similarities in their present-day languages.

These bold explorers travelled in sailing canoes from island to island, and must have been skilled navigators. According to DNA evidence, as the canoes travelled via the islands of Indonesia and New Guinea, some of the local people joined the travellers. By about 1,800 years ago the migrant explorers had settled Fiji, Tonga, Samoa, Cook Islands, Tahiti, and other island groups. All these island groups are known in tradition as Hawaiki. A French explorer later gave these islands the name 'Polynesia'.

New Zealand was the last place in the world to be settled by humans, somewhere between 1,000 and 700 years ago. Their descendants are Maori. According to analysis of how much mtDNA diversity there is among women of known Maori descent, about 170-230 women originally arrived in New Zealand. We don't know how many men, but it was probably a similar number. The gene mix has been changing as a result of intermarriage with later migrants. Among Maori men who have been tested so far, about half have Y-chromosomes identical to those found in European men.

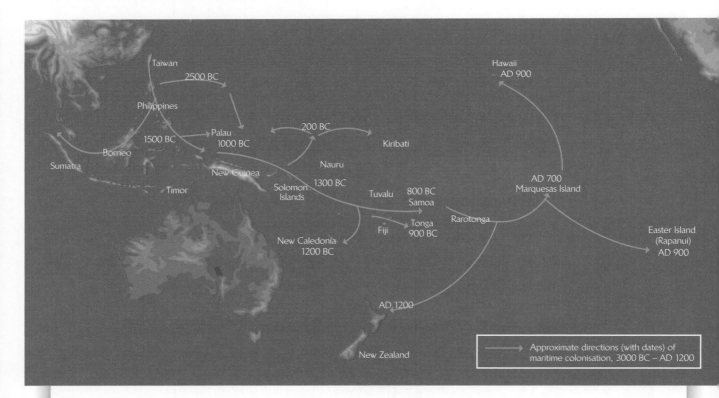

1 About when did people first arrive in Australia? _____

2 Where did Pacific people originate from 4,500 years ago?

_____

_____

_____

3 Using the map, name nine island groups included under the heading 'Polynesia'.

_____

_____

_____

_____

4 About what dates did people first arrive in Samoa and Tonga? _____

5 Over the past 1,000 years, what were the two main origins of genes in present-day Maori?

_____

_____

_____

6 What is the evidence for Taiwan being one place that Pacific people originally came from?

_____

_____

_____

_____

_____

CHALLENGE

ISBN: 9780170214667

1 Complete the definitions below by writing three words in each of the gaps.

Habitat means the p_____ o_____ a_____ where a particular species lives or a whole community lives. **Community** means a_____ l_____ t_____ in a particular place.

2 Jessica and Ben are given the task of finding which insects and other small life live in and on the patch of soil outside the science room. They are given information and equipment to catch the animals in two different ways (A and B), and then take them inside to identify using a microscope.

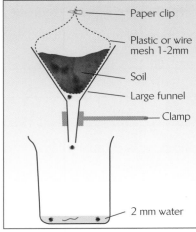

A Tullgren funnel, for catching soil and leaf-litter animals.

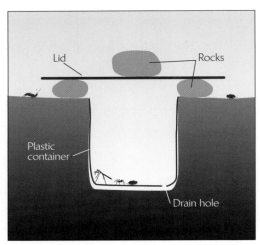

B Pitfall, to be used outdoors.

Compare and contrast these two methods. Describe some practical advantages and disadvantages of each method – including which kinds of animals it is likely to catch, or fail to catch. Mention any practical advantages, plus any difficulties of getting it to work.

A Tullgren funnel

_____
_____
_____
_____
_____

B Pitfall

_____
_____
_____
_____
_____

Ecology

SP2 Unit 2.12–2.14

Date for completion: / /

Parent sig: _____

Teacher sig: _____

Life Science

Ecology is the study of the relationships between living organisms (plants and animals), and between organisms and their environment.

1 Here are some ecology terms. Match each word to its short description: *community, ecosystem, environment, habitat, population, species.*

Short description	Word
A group of organisms that can breed with each other.	
The place where a type of organism normally lives.	
The physical conditions that affect living organisms.	
A group of plants and animals living and interacting with each other.	
Organisms of the same species living in the same area.	
Living and non-living linked together.	

2 Complete the following sentences, choosing from these words: *herbivores, Sun, carnivores, webs, omnivores, producers, photosynthesis.*

All organisms need to gain energy in order to live. Plants do this through

_____ (1), which turns the Sun's energy into food energy (chemical

energy). Animals that are _____ (2) eat plants in order to gain

energy. Animals like lions that eat only meat are called _____ (3),

whereas animals that eat both plants and animals are called _____ (4).

The patterns of who eats whom in a community can be summed up in food chains and food

_____ (5). Almost all food chains start with plants that make their own

food. These organisms are generally known as _____ (6), and they get

their energy from the _____ (7).

ISBN: 9780170214667

Date for completion: / /

Parent sig: _____
Teacher sig: _____

SP2 Unit 2.12/2.13

Life Science

Here are some of the organisms found living in or near a passion-fruit vine:

- aphids and caterpillars feed on the leaves of the passion-fruit vine
- moths and butterflies feed on the nectar in the flowers
- caterpillars are eaten by sparrows
- ladybirds feed on aphids
- sparrows eat butterflies and spiders
- spiders catch and eat ladybirds and moths
- cats eat sparrows
- owls (ruru) catch and eat sparrows and moths

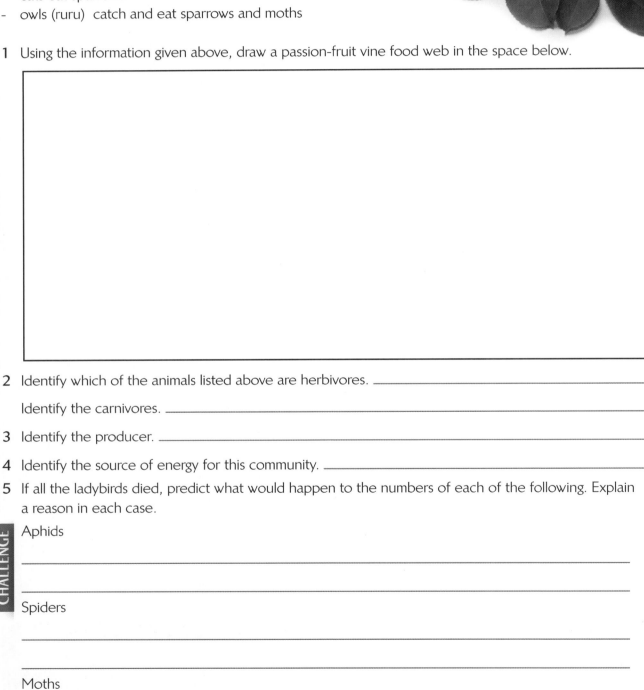

1 Using the information given above, draw a passion-fruit vine food web in the space below.

2 Identify which of the animals listed above are herbivores. _____

Identify the carnivores. _____

3 Identify the producer. _____

4 Identify the source of energy for this community. _____

5 If all the ladybirds died, predict what would happen to the numbers of each of the following. Explain a reason in each case.

CHALLENGE

Aphids

_____

_____

Spiders

_____

_____

Moths

_____

_____

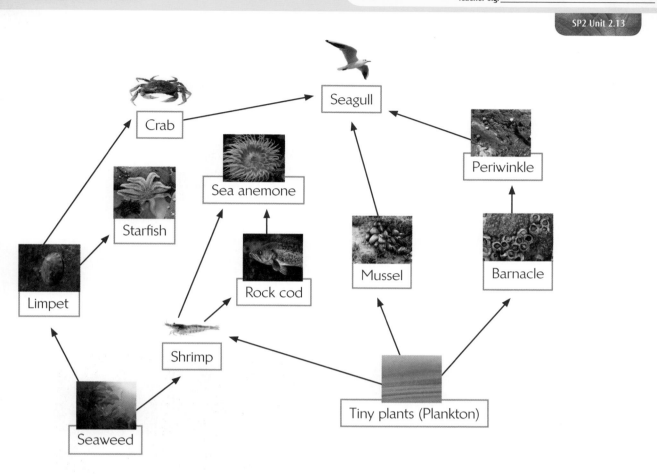

1 Name a carnivore in this rocky shore food web. _____

2 Name four herbivores in this food web.

_____

_____

3 Name two competitors, and the food they are competing for.

    competitors   _____

    food competed for   _____

4 Write three food chains that are part of the above food web. Each chain must start with a producer.

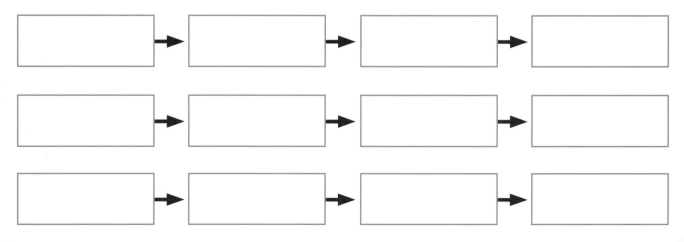

ISBN: 9780170214667

Date for completion: / /
Parent sig: _____
Teacher sig: _____

1 Match each relationship word to its description by drawing a linking line.

Type of relationship	Description
Mutualism	Where one organism benefits and the other is not affected.
Commensalism	Where one organism benefits and the other is harmed.
Predation	Where both organisms benefit by living together and depend on each other for their survival.
Competition	Where one organism feeds on the other to get its food.
Parasitism	When one organism is competing with the other for the same food supply.

2 Describe how the buffalo benefits from its partnership with the bird.

_____

_____

Describe how the bird is benefiting from this partnership.

_____

_____

Give a technical word for this kind of partnership.

_____

Community overview

Date for completion: / /

Parent sig: _____
Teacher sig: _____

SP2 Unit 2.13/2.14

Write each of these words in its correct place in this structured overview: *consumers, herbivores, decomposers, living things, omnivores, producers, carnivores.*

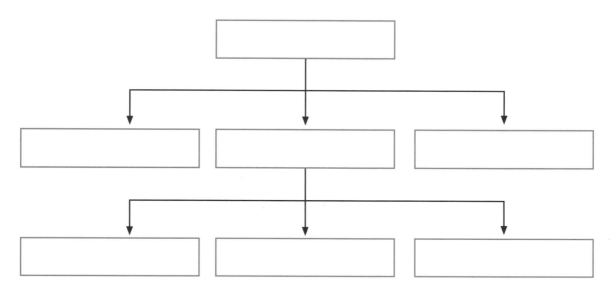

Energy in food chains

Date for completion: / /

Parent sig: _____
Teacher sig: _____

SP2 Unit 2.13/2.14

1 Complete these sentences, using the words from this list: *food, starch, heat, transferred, Sun, energy, sugars.*

Energy in almost every natural community starts with the _____.

Plants convert some of this light energy into _____ and _____ through a process called photosynthesis. At each trophic level, organisms use some of that _____ for day to day living. The energy stored in their _____ is released by respiration but eventually it is lost as _____. Generally, less than 10% of the energy is _____ at each step in the food chain.

ISBN: 9780170214667

2 Write each of these words in its correct box in the diagram below: *producers, third level consumers, herbivores (primary consumers), second level consumers, decomposers.*

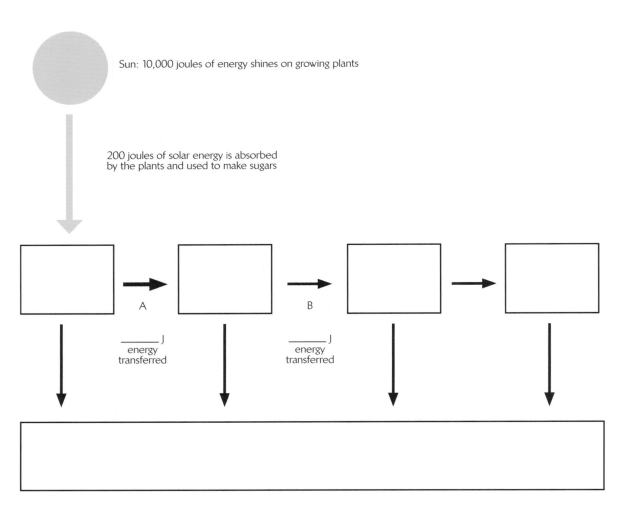

Sun: 10,000 joules of energy shines on growing plants

200 joules of solar energy is absorbed by the plants and used to make sugars

A

B

_____ J
energy transferred

_____ J
energy transferred

3 Next to arrows A and B above, write your realistic estimate of the number of joules transferred at these steps.

4 Write the name of any one living thing in each of the four smaller boxes above, if it matches that particular trophic level.

28 Carbon cycle

Date for completion: / /

Parent sig: _____
Teacher sig: _____

SP2 Unit 2.15

Answer the following, based on the carbon cycle diagram on the next page.

1 Name four processes that release carbon dioxide into the air.

_____
_____

2 Name two processes that take carbon dioxide from the air.

_____
_____

3 For the past 100 years, increased amounts of carbon dioxide in the air are almost certainly causing Earth to heat up. What is this increased heating known as?

_____

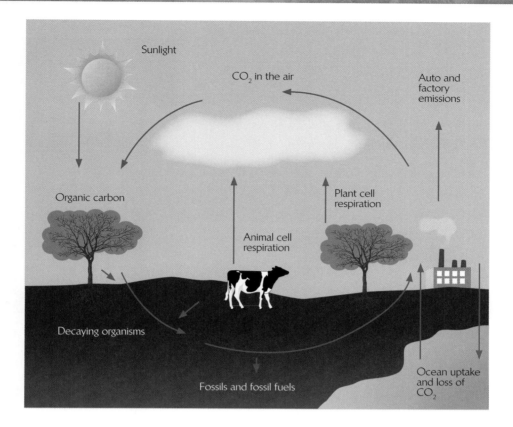

Sunlight

CO₂ in the air

Auto and factory emissions

Organic carbon

Plant cell respiration

Animal cell respiration

Decaying organisms

Fossils and fossil fuels

Ocean uptake and loss of CO₂

29 Biological control

SP2 Unit 2.16

Date for completion: / /   Parent sig: _____
Teacher sig: _____

The gum leaf skeletoniser (GLS) is an Australian moth that has the potential to defoliate thousands of hectares of eucalyptus plantations throughout New Zealand.

Entomologist Dr Lisa Berndt says the young caterpillars eat the green parts of gum tree leaves, avoiding the veins, which results in a skeleton-like appearance to the leaf. 'It's a very serious threat,' she says. 'Repeated attack by this pest can slow the tree's growth, or even kill individual trees. Outbreaks occur regularly in Australian eucalypt forests.'

Initially found in Mt Maunganui in the 1990s, the GLS was eradicated there by the Ministry of Agriculture and Forestry (MAF). However, in 2001 it was discovered in Auckland, and had already spread too far for eradication to be feasible. It is now widespread in the Auckland region, Waikato, Coromandel and Bay of Plenty, and is likely to continue spreading.

'Gum leaf skeletoniser is also hazardous to people,' says Lisa. 'The long hairs of the caterpillar inject venom into human skin on contact, which can cause a painful skin irritation.'

Lisa and her team have been developing a biological control to help manage this pest. The most suitable biological control agent was found to be an Australian parasitic wasp, *Cotesia urabae*. These lay their eggs in GLS caterpillars and the young wasp grubs then eat the caterpillar alive. This species specifically targets the GLS caterpillar. It is sourced from Tasmania, so will be adapted to a climate similar to New Zealand's.

After three years of research into the safety of the wasp's release into New Zealand, and consultation with a wide range of industry and community groups, the Environmental Risk Management Authority considered all the potential effects of the parasitic wasp to the environment, human health, the economy, local communities and Maori cultural values. Overall the benefits of release of the wasp outweighed any possible adverse effects.

Continued over

ISBN: 9780170214667

'Our research has shown that the wasp poses no significant risk to native insects in New Zealand, and it has no known effects on humans,' says Lisa. 'Biological control has been a very effective and safe means of control for other eucalyptus pests in New Zealand. We hope this new agent will be successful on GLS because it is the only sustainable alternative to applying chemical insecticides to our forests.'

1 Summarise why the GLS is a problem. (One main reason, and one secondary reason.)

_____

_____

_____

2 To what areas has it spread?

_____

_____

3 In 20-40 words, describe what the GLS biological control 'agent' is, and how it works.

_____

_____

_____

_____

_____

_____

_____

4 What is the alternative to biological control of the GLS?

_____

5 List three factors that make the introduced wasp a suitable biological control agent in this case.

_____

_____

_____

Nuts to you

Brazil nuts come from just one species of rainforest tree. Every Brazil nut you've ever eaten has been collected from the Amazon jungle. In the past, people who tried to farm these trees for their valuable crop were disappointed to find that the trees almost never produce seed. Something was missing.

These trees need intact rainforest. This is because Brazil nut trees need their natural undamaged ecosystem in order for their flowers to be pollinated, and that ecosystem includes several kinds of wild forest bees.

These wild bees don't feed on orchid flowers. Instead, the male bees crawl all over the flowers, collecting chemicals from special flower glands. This is the only way males are able to make the sex attractant necesssary in order to attract bee females. In their scramble, the males spread orchid pollen from one flower to another.

Different orchid species produce unique mixes of scent chemicals, and the female bees prefer males who do not smell of other kinds of flower. This means that males typically visit only one species of orchid, which increases the bee's chances of mating successfully. This increases the orchid flower's chance of pollination, and also prevents cross-pollination between different orchid species.

What has all this got to do with Brazil nuts? It turns out that female wild bees are the only insects that can pollinate the flowers of Brazil nut trees. Without intact rainforest and the right kind of climate, there can be no orchids. Without orchids, no sex for male bees. Without female bees there would be no pollination, and no Brazil nuts.

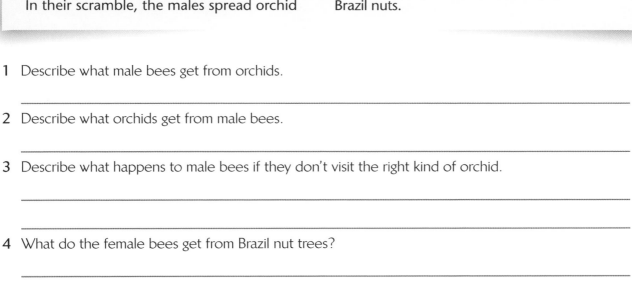

Life Science

1 Describe what male bees get from orchids.

_____

2 Describe what orchids get from male bees.

_____

3 Describe what happens to male bees if they don't visit the right kind of orchid.

_____

_____

4 What do the female bees get from Brazil nut trees?

_____

5 What do the Brazil nut trees get from female bees?

_____

ISBN: 9780170214667

CHALLENGE

6 The Brazil nut information adds up to one small example of how living things are inter-dependent. Next to each of the six arrows below, sum up in 2-4 words how that plant or animal is affected by that part of the overall arrangement.

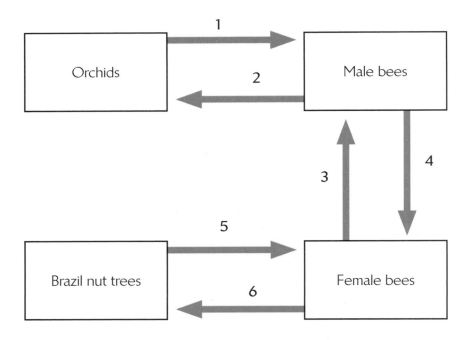

31 Bacteria

Date for completion: / /

Parent sig: _____
Teacher sig: _____

SP2 Unit 2.21

1 Label the diagram below. Choose six from this list of ten terms: *cytoplasm, cell wall, cell membrane, chloroplast, binary fission, capsule, pathogen, flagellum, nucleus, DNA.*

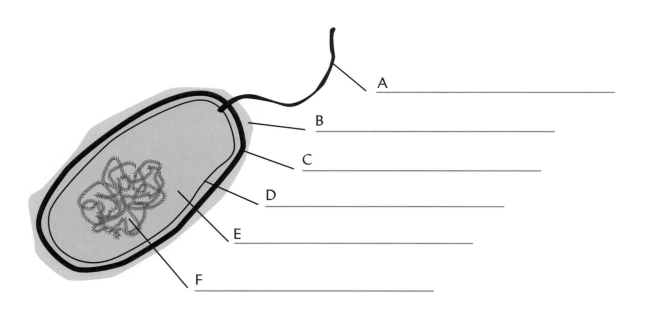

2 Circle six words in this word find by using labels from the diagram in question **1**.

```
G  S  D  Y  R  N  V  F  W  D  Q  X
F  S  P  J  O  D  K  V  M  O  R  R
T  Y  M  R  N  D  V  B  B  H  T  U
D  I  J  A  Y  O  K  Q  T  R  N  N
R  P  D  D  C  C  A  P  S  U  L  E
C  E  L  L  M  E  M  B  R  A  N  E
P  N  M  S  A  L  P  O  T  Y  C  M
S  N  H  C  N  D  W  T  D  R  D  E
B  C  O  U  S  Q  G  K  Q  X  R  W
L  S  J  F  L  A  G  E  L  L  U  M
Z  R  I  V  X  S  P  H  T  U  U  K
C  E  L  L  W  A  L  L  L  I  J  T  S
```

3 Bacteria use the process of **binary fission** to reproduce. They use **extra-cellular digestion** to get food from their surroundings.

In 4 – 10 words, explain what each of these words mean:

Binary means _____

Fission means _____

Digestion means _____

Extra-cellular means _____

Bacterial reproduction

Date for completion: / /

Parent sig: _____

Teacher sig: _____

SP2 Unit 2.21

Bacteria reproduce non-sexually by **binary fission**. Some types of bacteria can divide every 10 minutes, while others can take 10 hours to divide. *E. coli* is a type of bacteria that can cause food poisoning in humans. They cause cramps, vomiting and diarrhoea. Yuck! They can divide about every 20 minutes.

Imagine that there are 10 *E. coli* bacteria present on the corner of a muffin because it had touched the counter top. The data on the next page shows how many bacteria would be present on the muffin after four hours, if growth continued at the same rate.

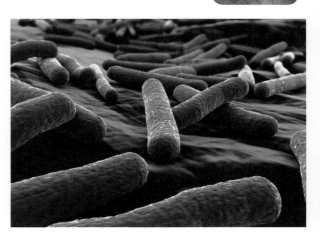

ISBN: 9780170214667

Time (min)	Number of bacteria
0	10
20	20
40	
60	80
80	160
100	320
120	
140	1280
160	2560
180	5120
200	
220	
240	40960

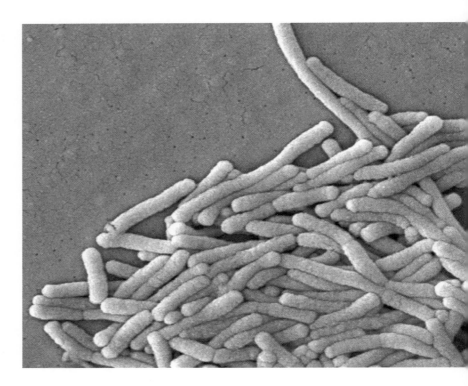

1 Fill in the missing numbers, based on doubling every 20 minutes.
2 Using the data above, draw a graph with time on the horizontal (x) axis, and the bacteria population on the side (y) axis. Allow for 200 minutes. Give your graph a title.

ISBN: 9780170214667

3 Using your graph, predict approximately how many bacteria would be present after 170 minutes.

_____

4 Estimate how many bacteria would be present after 200 minutes by following (extrapolating) the line on your graph. Write this number on your graph.

5 Calculate how long it would take this population of _E. coli_ to reach over 1 000 000.

_____

6 Would this population reproduce differently if the muffin was kept in the fridge? Explain.

_____

_____

Date for completion: / /

Parent sig: _____
Teacher sig: _____

SP2 Unit 2.21/2.22

Fill in the blanks to complete the sentences below. Use the answers to complete the red panel below.

Bacteria reproduce by a process called _____ (1: two words). Many kinds

of bacteria and fungi are called _____(2), because they feed off living

things and can cause disease. A mass of fungal hyphae is called a _____(3)

Some types of bacteria and fungi are called _____(4), which

means means that they feed off dead and decaying matter. Fungi reproduce by releasing

_____(5). A fungus _____ (6) is also called a

spore case because it produces and holds the spores until they are mature and can be released.

Bacteria in the guts of plant-eating animals help to _____ (7) food. The

_____ (8) is the outermost layer of a bacterial cell.

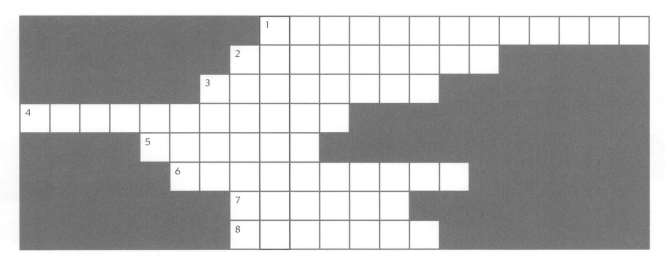

ISBN: 9780170214667

34 False!

Date for completion: / /
Parent sig: _____
Teacher sig: _____

SP2 Unit 2.21/2.22

Each one of these statements is incorrect in some way. Rewrite each of the following sentences, making at least one correction to each so that the statement becomes true.

1 A bacterium's genetic material is found inside a nucleus.
2 Bacteria reproduce by binary fusion.
3 Bacteria feed by using their spores.
4 Bacteria can be saprophytes or paraffins.
5 If the conditions are right, bacteria can reproduce every minute.
6 Bacteria feed by releasing hormones onto their food to help break it down into smaller particles.
7 Athlete's foot is an example of an infection caused by bacteria.
8 Most bacteria are harmful, and not many kinds are useful.

1 _____
_____

2 _____
_____

3 _____
_____

4 _____
_____

5 _____
_____

6 _____
_____

7 _____
_____

8 _____
_____

35 Fungi

Date for completion: / /
Parent sig: _____
Teacher sig: _____

SP2 Unit 2.22

Different fungi have different ecological relationships with green plants. Three types are listed in the table on the next page, but the sentences are not matched up. Choose sentences from each column (A, B, C), and write them to make three short paragraphs that describe each type of relationship. Write in the spaces below the table. Also identify which picture is most closely matched to each paragraph.

A	B	C
Some fungi are mutualists.	They get their food from the cells of living hosts.	They then absorb nutrients from dead remains.
Some fungi are parasites.	They digest dead plants and animals.	In return the hosts provide sugars for the fungi.
Some fungi are decomposers.	They can help their plant host get water and nutrients from soil.	The hosts are harmed by the fungi.

_____

_____

_____

Picture: _____

_____

_____

_____

Picture: _____

_____

_____

_____

Picture: _____

This fungus feeds by digesting dead wood.

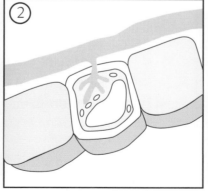

This fungus is getting its food from a plant cell, and is in turn helping the plant absorb water and nutrients from the soil.

Leaf spot fungus.

ISBN: 9780170214667

Life Science

Use the following organiser to compare and contrast bacteria and fungi.

Bacteria	Fungi

Similarities

Differences

reproduction

movement

growth

Date for completion: / /

Parent sig: _____
Teacher sig: _____

SP2 Unit 2.23

Life Science

1 Penicillin was discovered by Alexander _____ in 1928.

Apart from penicillin, two other antibiotics are _____ and

_____ . Antibiotics are effective against bacterial infections, but not against

_____ diseases such as _____ .

2 The graph shows the numbers of antibiotic-resistant MRSA infections in New Zealand. Use the information to answer questions A to D.

A Estimate by what factor (i.e. by how many times) the number of MRSA cases increased from 1994 to 2002.

_____

B Estimate the number of cases in 1999.

_____

C If action to deal with MRSA had not been taken in 2000 - 2001, estimate how many cases there might have been in 2005.

_____

D Explain in 20 - 40 words how bacteria become resistant to antibiotics.

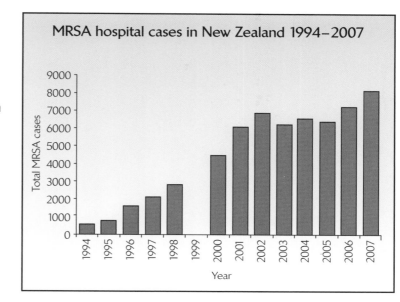

MRSA hospital cases in New Zealand 1994–2007

CHALLENGE

_____

_____

_____

Date for completion: / /

Parent sig: _____
Teacher sig: _____

SP2 Unit 2.21/2.23

1 I am a micro-organism that feeds on living hosts. _____

2 I am a chemical secreted by fungi to break down food. _____

3 I am a micro-organism that causes disease. _____

4 I am the method of reproduction in bacteria. _____

5 I am the part of some bacteria that provides motion. _____

6 I am the part of fungi that make them look fuzzy. _____

7 I am a solution that kills micro-organisms on surfaces. _____

8 I am the part of a fungus that is responsible for feeding. _____

9 I am the part of bacteria that make them appear shiny. _____

10 I am a fungus that ferments sugar into alcohol and CO_2. _____

ISBN: 9780170214667

Group the words below to create a mind map of the main information in units 31 to 38.
A mind map consists of lots of little bubble diagrams linked together by arrows. You can add pictures to help you remember the words.

binary fission	saprophyte	pathogen	enzyme	capsule
flagellum	recycling	fermenting	sporangium	hyphae
cell wall	spore	penicillin	mould	yeast
cell membrane	antibiotic	aid digestion	resistance	mycelium

Life Science

40 Digestion word find

Date for completion:	/ /	Parent sig: _____
		Teacher sig: _____

SP2 Unit 2.24

Circle each of the following words where it appears in the grid below.

COLON CARBOHYDRATES TONGUE INTESTINE
ENZYME OESOPHAGUS TEETH AMINOACIDS
STOMACH FAT BILE
PROTEIN PANCREAS SALIVA

```
                        B  L  Q
                  D  D  I  L  K  J  X  B  E
               V  I  Y  O  E  N  Z  Y  M  E  H  X  Y
            N  K  Q  E  N  V  L  G  Y  O  C  M  E  Y  U  T  P
         A  Y  L  H  S  N  T  V  L  X  N  F  G  N  Z  C  T  B  O
      Z  T  T  D  A  W  R  E  E  E  O  B  F  M  O  E  A  P  D  E  Q
      B  H  L  E  K  R  V  E  L  S  N  Q  B  D  S  F  R  I  W  S  I
   G  R  L  R  T  U  I  R  T  A  L  T  A  B  D  W  T  B  X  C  O  X  L
   V  K  C  A  V  L  K  R  H  A  L  B  I  I  A  T  L  O  O  O  P  X  V
O  R  N  Q  V  W  A  K  X  A  H  Y  T  Y  N  W  D  A  H  X  L  H  H  A  H
B  A  U  N  X  S  B  N  E  X  K  M  O  B  M  E  L  L  Y  Q  O  A  V  B  G
P  O  R  V  N  C  T  H  I  J  C  G  R  H  M  N  X  B  D  B  N  G  A  G  J
R  C  E  Q  R  G  J  K  O  N  K  R  X  R  U  L  O  B  R  R  Z  U  O  O  V  T
P  V  T  M  T  A  J  Z  D  M  N  B  Z  O  W  O  T  X  U  A  B  E  S  D  G  B  S
P  Z  J  G  A  X  F  N  Z  C  A  Q  Y  A  F  V  C  C  X  T  U  Z  X  O  M  G  L
B  L  Z  E  C  G  H  D  Y  C  C  W  M  M  V  R  H  S  E  P  T  S  A  K  P
Q  X  W  H  H  H  R  R  W  G  J  H  I  H  U  K  B  A  S  Q  Z  A  A  E  W
W  Y  B  I  P  T  M  A  O  O  R  P  N  H  P  D  T  L  A  F  L  E  D  I  K
   P  Z  N  D  M  Q  N  D  Q  E  D  O  R  D  I  U  I  L  R  M  N  R  O
   K  E  H  S  P  X  X  P  J  W  A  A  I  D  C  E  V  P  V  S  N  U  P
   N  K  P  B  N  R  R  O  I  G  C  X  Z  Y  H  A  E  N  N  Y  V
   R  C  R  Z  I  D  O  K  V  V  I  F  G  I  V  U  C  S  T  G  B
      N  W  I  R  L  T  V  T  I  D  V  N  L  G  B  Z  J  J  R
      D  B  H  S  E  N  T  X  R  O  E  N  E  D  J  Y  G
         B  M  I  L  Y  K  P  N  O  T  U  Y  S
            N  B  Q  S  B  T  H  P  O
                  L  C  S
```

41 Digestion summary

Date for completion:	/ /	Parent sig: _____
		Teacher sig: _____

SP2 Unit 2.24/3.15

1   Complete this table using words from this list: *starch, lipase, stomach, protein, pancreas, lipids (fats), saliva, protease, amylase.*

Enzyme	Where it is made	What it digests

ISBN: 9780170214667

2 Summarise the action of digestive enzymes by filling in the missing words:

Digestion of lipids (fats)

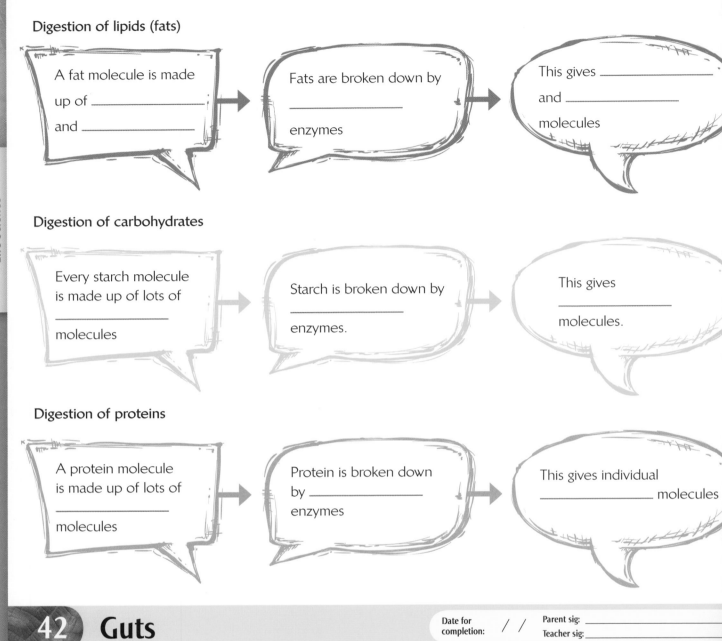

A fat molecule is made up of _____ and _____

Fats are broken down by _____ enzymes

This gives _____ and _____ molecules

Digestion of carbohydrates

Every starch molecule is made up of lots of _____ molecules

Starch is broken down by _____ enzymes.

This gives _____ molecules.

Digestion of proteins

A protein molecule is made up of lots of _____ molecules

Protein is broken down by _____ enzymes

This gives individual _____ molecules

42 Guts

Date for completion: / /

Parent sig: _____
Teacher sig: _____

SP2 Unit 2.24

1 Complete this table which gives a short summary of four structures in the digestive system. The first row has been done for you. The next three are not listed in order.

Structure	Action	Function
Teeth	Break up food into smaller pieces.	In order to make the pieces easier to swallow, and to increase surface area for the enzymes.
Small intestine		
Pancreas		
Salivary glands		

ISBN: 9780170214667

2 Label the structures **A** to **K**.

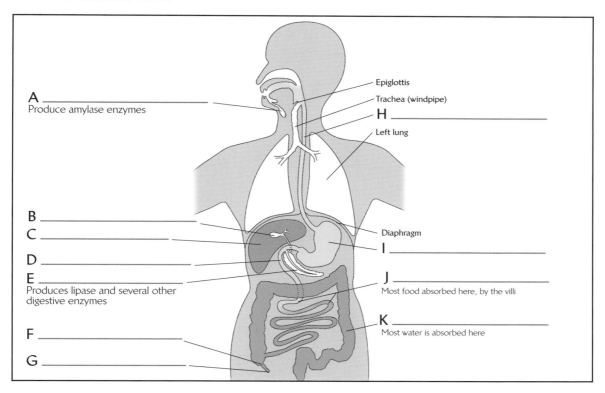

A _____
Produce amylase enzymes

Epiglottis

Trachea (windpipe)

H _____

Left lung

B _____
C _____
D _____
E _____
Produces lipase and several other
digestive enzymes

F _____

G _____

Diaphragm

I _____

J _____
Most food absorbed here, by the villi

K _____
Most water is absorbed here

Life Science

CHALLENGE

3 Animals like cows, sheep and deer have a special stomach section known as the **rumen**. Describe what is special about the type of digestion that happens in a rumen.

_____

_____

_____

_____

43 **Body balance**

Date for completion: / /

Parent sig: _____
Teacher sig: _____

SP2 Unit 2.25

1 Complete the following sentences, choosing from this word list: *opposite, negative, sugar, evaporation, feedback, thermostat, normal, reduces, increase.*

Your body has control systems that keep your body temperature and blood

_____ nearly constant. These systems work through a process of negative

_____. For example, when controlling your body temperature, your brain

detects overheating, then sends nerve signals that _____ the rate that you

sweat. The _____ from your skin then has a cooling effect. When your

temperature returns to its _____ set level of

37° C, sweating then _____ again . This kind of system is described as

_____ feedback because it has the _____ effect

to the original change. A _____ is an example of a technical device that

uses negative feedback.

ISBN: 9780170214667

2 This flow diagram shows the negative feedback changes that help return your blood sugar to a proper level after a meal. Short statements that belong in each of the boxes are listed here, but out of sequence. Decide which statement goes where, and write each in its correct box.

blood sugar level falls *less insulin is produced now*
blood sugar level rises *insulin tells the liver to store sugar*
starch is digested to sugar *the pancreas releases insulin*

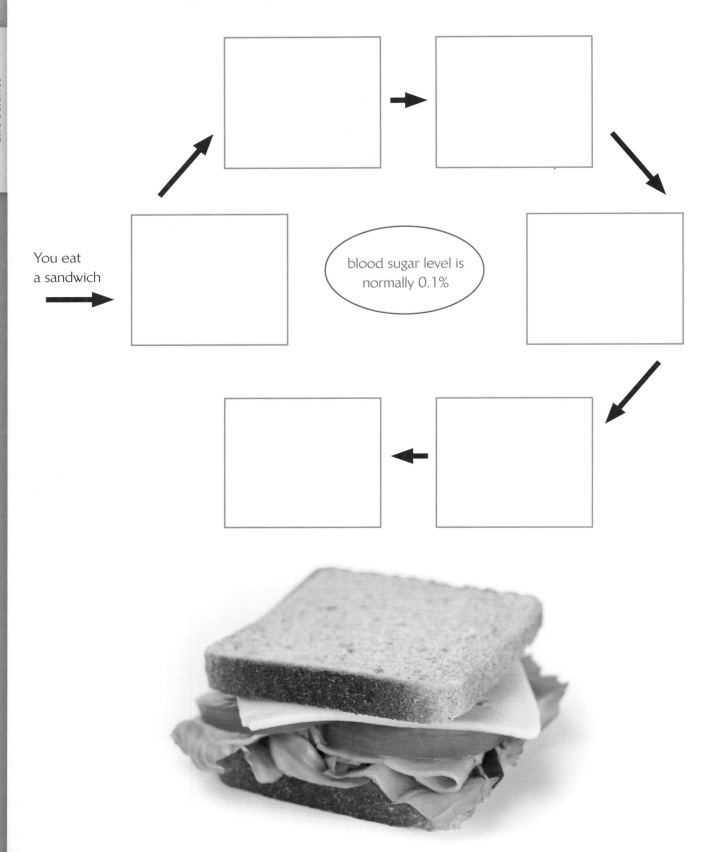

You eat a sandwich

blood sugar level is normally 0.1%

SP2 Unit 2.27

Life Science

Read the following article, then use the information to answer questions 1-6.

White blood cells are our main form of defense against infections. There are many different types of white blood cells. Granulocytes swallow bacteria and viruses, and also destroy cells that have died or become infected or are in the process of developing into cancer cells.

Another type of white blood cell is a lymphocyte. These lymphocytes 'recognise' foreign material known as antigens, then start a process which leads to the production of antibodies. These antibodies are special protein molecules that bind to the invading viruses or bacteria, and prevent the invaders from reproducing.

Lymphocytes can also 'remember', so if a type of bacteria or virus has entered the body before, the immune system can quickly make the antibodies to help destroy it. Remembering is an important feature of immunity, as this is how you become resistant to a particular type of attacker. Your body recognises invaders that have been encountered before, and destroys them before you get sick.

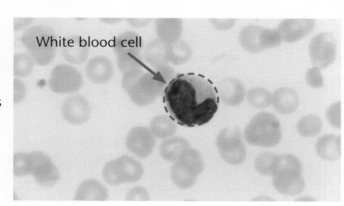

White blood cell

1 Name two types of white blood cells.

_____

2 Describe two ways in which the white blood cells fight off infections.

_____

_____

3 Which type of cell helps produce antibodies?   _____

4 Which type of white blood cell engulfs bacteria and viruses?

_____

5 Explain the difference between antigens and antibodies.

_____

_____

_____

CHALLENGE

6 The number of white blood cells we have changes depending on our health. Predict in what situations you would be likely to have unusually high numbers of white blood cells. Explain why.

_____

_____

_____

_____

ISBN: 9780170214667

Chemical Science

1 Particle puzzle

Date for completion: / /

Parent sig: _____
Teacher sig: _____

SP2 General

A Revision task: use the clues to help you unscramble words 1 to 14.

1 PAICERTLS

6		17				26		

2 TCARTAT

					13		1

3 LEECUMOL

	18					8	

4 TONORSP

9	24	21				

5 ECHAGR

	2	23		25	

6 XGOYNE

		14		3	

7 SEAGS

		4		

8 NELROTCES

		22			5			

9 DLOME

		7		

10 SUNNEOTR

10	19			16			

11 WARET

		11		

12 SUNUCLE

				12		

13 NAD

20		

14 NIOS

	15		

Clues:

1 A general word for atoms and molecules.
2 Opposite-charged particles ___ each other.
3 Two or more atoms joined together.
4 Positively charged particles in atoms.
5 An electric _____ can be positive or negative.
6 Gas making up almost 21% of the air.
7 The particles in _____ are far apart and move quickly.
8 The smallest particles in atoms.

9 A way of representing something that is invisible.
10 Found together with protons in the nucleus of most atoms.
11 Each molecule of this substance has two hydrogen atoms and an oxygen atom.
12 The central part of an atom.
13 A large complex molecule found in all living things.
14 Atoms that have lost or gained electrons are called _____ .

ISBN: 9780170214667

B A neutron walks into the dairy to get a bag of lollies. It asks the shopkeeper, 'how much?' To find the answer, use numbered letters from your answer to part A.

1 2 3	4 2 5 6	K
	7 8 9 10 11	12 13 14 15

F	
16 17	

14 18 19	20 21	22 2 23 24 25 26

Elements revision

Date for completion: / /

Parent sig: _____

Teacher sig: _____

SP2 Unit 3.1–3.20

Chemical Science

Everything around you is made of chemical elements. Revise the following chemical elements. If you have not learned their symbols before now, this is a good time to start. Use the periodic table (page 160) to complete this summary.

	Symbol	Full name	Any one fact about it
1	H		Explosive gas
2	He		
3	Li		The lightest metal
4	Be		A light poisonous metal
5			A rare micro-nutrient
6		Carbon	
7			Makes up 78% of the air
8			
9			In your teeth enamel
10	Ne		
11			An element in food salt
12			This metal burns easily
13		Aluminium	
14			A major element in most rocks
15			Very dangerous in pure element form
16			Yellow powder around volcanic vents
17			
18			Gas making up almost 1% of the air
19			Dangerous metal, but an essential nutrient
20			
26			
28			A metal used in some coins
	Cu		
30			A metal used for rust-proofing
	Ag		Used in some jewellery
		Iodine	Needed by your thyroid gland
	Au		
	Hg		
	Pb		
	U		

ISBN: 9780170214667

3 Atom revision

Date for completion: / / Parent sig: _____ Teacher sig: _____

SP2 Unit 3.1–3.20

1 In this atom diagram, state what is represented by:

the black dots _____

12 p _____

12n _____

2 What element is it? _____ Does the diagram represent an

atom or an ion? _____

3 Supply the missing words by writing **positive** or **negative**.

If an atom **loses** one or more electrons, it becomes a _____ ion. If an atom **gains** an

electron it becomes a _____ ion.

4 Write the formula for this element when it becomes an ion. _____

---

4 Metal compounds

Date for completion: / / Parent sig: _____ Teacher sig: _____

SP2 Unit 3.8

When aluminium (Al) comes into contact with oxygen (O), it forms a compound called aluminium oxide. The formula for aluminium oxide is Al_2O_3. There are reasons for the different numbers for Al and O.

One way to understand this is to imagine + and − charged particles. The overall charge is zero, as in this diagram of $MgCl_2$.

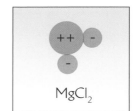

$MgCl_2$

Another way is to visualise atoms as puzzle pieces. Al is in Group 13 so has three outer electrons. Oxygen is in group 16, so has space for two more electrons in its outer shell. Of course atoms don't really look like puzzle pieces, but the method may help you understand the ratios in which they combine.

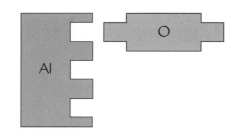

In order for Al and O to both have full outer shells, two aluminium pieces link with three oxygen puzzle pieces like this:

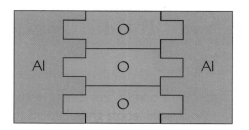

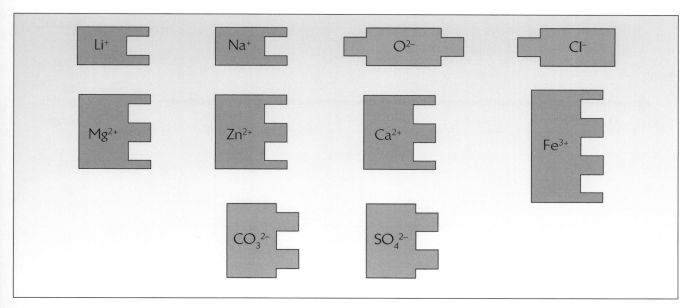

Now use this method to draw 'joined puzzle pieces' in each box.

Sodium sulfate	Magnesium oxide	Calcium carbonate
Zinc oxide	Lithium sulfate	Zinc carbonate
Magnesium sulfate	Sodium carbonate	Iron oxide

ISBN: 9780170214667

Chemical Science

A sulfur atom is represented by the symbol S. If a sulfur atom gains two electrons, it becomes a sulfide ion, symbol S^{2-}. (Some books write S^{-2}, but S^{2-} is preferred.)

1 Complete this table.

Atom symbol	Atom loses or gains how many electrons?	Ion symbol	Ion name
S	Gains 2		sulfide
O	Gains 2		oxide
Cl	Gains 1		
	Loses 1		sodium
K	Loses 1		potassium
Cu	Loses 2		
Mg		Mg^{2+}	
Zn		Zn^{2+}	
		Al^{3+}	
	Loses 1		hydrogen ion

2 The above table shows that metal atoms tend to _____ gain / lose _____ electrons, and non-metal atoms tend to _____ gain / lose _____ electrons.

3 Use the information in the 'ions and formulae' task to complete the following table. Make sure you memorise the chemical names of different ions.

Ion name	Ion symbol
hydrogen	
sodium	
copper	
oxide	
chloride	
ammonium	
hydroxide	
sulfate	
nitrate	
carbonate	
hydrogen carbonate ('bi-carbonate')	

4 Match each of these definitions (**A** to **H**) with one word chosen from this list: *carbonate, ion, sulfate, oxide, ionic, salt, covalent, compound.*

A An electrically charged atom, or group of atoms. _____

B A compound in which metal ions are combined with non-metal ions. _____

C Chemical bonds between opposite-charged ions. _____

D Chemical bonds that result from electron-sharing. _____

E A name for the CO_3^{2-} ionic group of atoms. _____

F A name for the SO_4^{2-} ionic group of atoms. _____

G A chemical substance that contains more than one element. _____

H A compound containing a metal ion and an oxygen ion. _____

6 Ions and formula revision

SP2 Unit 3.19

Date for completion: / /   Parent sig: _____ Teacher sig: _____

This table of ion charges (also known as a valency table) makes it possible to work out chemical formulae. When ions from the positive (blue) side of the table combine with ions from the negative (orange) side, the overall charge is always zero.

Electrical charge on the ion (names given here for poly-atomic ions only)					
3+	2+	1+	0	1-	2-
Al	Mg	Li	Inert gases like neon, helium and argon do not gain or lose electrons, so do not form ions and do not react.	Cl	O
	Ca	Na		I	S
	Zn	K		OH (hydroxide)	SO_4 (sulfate)
	Cu	NH_4 (ammonium)		HCO_3 (hydrogen carbonate)	CO_3 (carbonate)
				NO_3 (nitrate)	

Many reactions between acids, bases and metals result in compounds which are ionic. In the table below write the correct formula of compounds 1 – 20. But first, a reminder of what a formula represents.

The **letters** show which elements make up the compound. Metal atoms are written first.

$$Cu\ SO_4 \qquad MgCl_2$$

The **numbers underneath** show how many of each kind of atom make up one molecule of the compound. If no number is given, it means there is only one of that kind of atom.

Many reactions between acids, bases and metals result in compounds which are ionic. In the table below write the correct formula of compounds 1-20. But first, a reminder of what a formula represents.

	Name	Formula	Atoms *
1	Calcium oxide	CaO	2
2	Potassium chloride		
3	Zinc chloride		
4	Aluminium chloride		
5	Potassium sulfate		
6	Zinc sulfate		
7	Aluminium sulfate		
8	Lithium carbonate	Li_2CO_3	6
9	Calcium carbonate		

ISBN: 9780170214667

	Name	Formula	Atoms *
10	Calcium nitrate		
11	Sodium oxide		
12	Magnesium oxide		
13	Aluminium oxide		
14	Sodium hydroxide		
15	Copper hydroxide	$Cu(OH)_2$	5
16	Copper chloride		
17	Copper iodide		
18	Sodium hydrogen carbonate		
19	Calcium hydrogen carbonate	$Ca(HCO_3)_2$	
20	Ammonium hydroxide		

* This the total number of atoms shown in the formula, regardless of whether the substance has molecules or ions.

> If you struggle with this task, go to *SciencePlus 2* units 3.19 and 3.20 for more on compound naming rules and on chemical equations.

7 Equations introduced

Date for completion: / /

Parent sig: _____

Teacher sig: _____

SP2 Unit 3.1–3.20

> If you are not yet confident with simple chemical equations do this task before trying to balance equations.

Examples of chemical reactions are all around and inside us: cooking, rusting, burning, digesting food. A chemical equation is a convenient way of describing what happens in a chemical reaction.

What happens in most chemical reactions is that atoms change partners. The atoms do not vanish, and they do not change into other kinds of atom.

Chemical equations can be written either with words, or with chemical symbols. Either way, an equation can be simplified like this:

One example of a colourful reaction: if you add some lead nitrate solution to a watery solution of potassium iodide, a yellow substance will appear. Two compounds have reacted to make lead iodide, which is the yellow substance. All this can be summed up simply into a one line word equation.

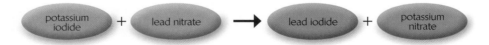

1 Crystals of solid sodium chloride (common salt) form when melted sodium meets chlorine gas.

A Name the reactants. _____

B Name the products. _____

C Write a word equation for the reaction. _____

_____

2 A piece of copper wire is put into silver nitrate solution. Crystals of shiny silver metal start to grow on the copper. Eventually all the copper reacts to form blue copper nitrate solution.

   A Name the reactants. _____

   B Name the products. _____

   C Write a word equation for the reaction. _____

   _____

3 In some cases two clear solutions react, and one of the products is a solid that forms a cloudy mixture. In these cases, the solid is called a **precipitate**. Copper sulfate solution and sodium hydroxide react to form a blue precipitate of copper hydroxide. Sodium sulfate solution is also formed.

   A Name the reactants. _____

   B Name the products. _____

   C Write a word equation for the reaction. _____

   _____

4 If you heat calcium carbonate, it breaks down to form CO_2 and calcium oxide.

   A Name the reactant. _____

   B Name the products. _____

   C Write a word equation for the reaction. _____

   _____

Equations balanced

Date for completion: / /   Parent sig: _____
                              Teacher sig: _____

SP2 Unit 3.20

A chemical equation sums up what happens in a reaction. Any equation can be written using words, or by using symbols and formulae. Example:

Words: magnesium + sulfuric acid $\longrightarrow$ magnesium sulfate   +   hydrogen
Formulae: Mg   + H_2SO_4 $\longrightarrow$ $MgSO_4$ + H_2

A correctly-written formula equation shows the true situation:
- no atoms are destroyed, no new atoms are made, no elements are changed.
- after a reaction there are exactly as many atoms of each kind as there were at the start.

Difficulties arise when an equation seems to show more atoms on one side than on the other. The unbalanced example below is wrong, because the extra H and Cl atoms can't magically appear from nowhere. In many situations, you need to balance an equation by writing a number in front of one or more of the formulae. Decide which. The goal is to write a balanced equation that shows exactly the same number of atoms of each kind before and after the reaction.

Unbalanced: Mg +   HCl $\longrightarrow$ $MgCl_2$ + H_2
Correctly balanced: Mg +   2 HCl $\longrightarrow$ $MgCl_2$ + H_2
(We can't balance by writing MgCl, because no such molecule exists.)

ISBN: 9780170214667

Equations 2 to 20 are not yet balanced. Balance each one by writing '2' (or 3 or 4 or 6) in front of some formulae, as needed.

Elements reacting

1    $2Ca$ + O_2 → $2CaO$

2   Na + O_2 → Na_2O

3   C + O_2 → CO

4   Al + O_2 → Al_2O_3

Bases plus acids

5   Na_2O + HCl → $NaCl$ + H_2O

6    $NaOH$ + H_2SO_4 → Na_2SO_4 + H_2O

7   KOH + H_2SO_4 → K_2SO_4 + H_2O

8    $Ca(OH)_2$ + HCl → $CaCl_2$ + H_2O

9   ZnO + HCl → $ZnCl_2$ + H_2O

10    $Mg(OH)_2$ + HCl → $MgCl_2$ + H_2O

Carbonates plus acids

11    $CaCO_3$ + HCl → $CaCl_2$ + H_2O + CO_2

12    $MgCO_3$ + HCl → $MgCl_2$ + H_2O + CO_2

13    $ZnCO_3$ + HCl → $ZnCl_2$ + H_2O + CO_2

Metals plus acids

14   Zn + HCl → $ZnCl_2$ + H_2

15   Li + H_2SO_4 → Li_2SO_4 + H_2

16   Al + HCl → $AlCl_3$ + H_2

Some other reactions

CHALLENGE

17   H_2O_2 → H_2O + O_2 (the breakdown of hydrogen peroxide)

18    $C_6H_{12}O_6$ → C_2H_5OH + CO_2 (fermentation simplified)

19   H_2O + CO_2 → $C_6H_{12}O_6$ + O_2 (photosynthesis simplified)

20    $C_6H_{12}O_6$ + O_2 → H_2O + CO_2 (cell respiration simplified)

Date for completion: / /
Parent sig: _____
Teacher sig: _____

SP2 Unit 3.1/3.2

Match the descriptions in the left column to one of terms listed below. Write these terms in the correct place.
pH14; pH 7; hydrogen; pH 1; hydroxide; pH 5; indicator; pH; pH 9; logarithmic.

Description	Matching term
Strong acid	
Strong base	
Short for 'potential of hydrogen'	
Weak acid	
Neutral	
Weak base	
Ion found in all acids	
Ion found in most bases	
A scale that goes in multiples of ten	
Changes colour according to pH	

Date for completion: / /
Parent sig: _____
Teacher sig: _____

SP2 Unit 3.1/3.2

Classify the words and symbols in the middle, by writing each in either the Acid column or the Base column. (If there is not enough information to decide on an item, circle it and leave it in the central column.)

Acid	Acid or base?	Base
	pH 8	
	sour	
	fizzy drink	
	pH 6	
	sodium hydroxide	
	pH 12	
	red	
	pH 1	
	blue	
	KOH	
	toothpaste	
	H_2SO_4	
	soap	
	fruit juice	
	oven cleaner	
	alkali	
	HCl	
	$Mg(OH)_2$	

Chemical Science

ISBN: 9780170214667

Date for completion: / /

Parent sig: _____

Teacher sig: _____

Sophie took a sample of compound A, added it to solution B, and found that gas C bubbled off. She filtered the resulting solution D, then gently heated the filtrate until almost all the water had evaporated. She left the solution overnight and the next day found blue crystals had formed.

Compound A is a green powder.
Compound B is a solution that turns blue litmus red.
Solution D has no affect on litmus paper.
The blue crystals were copper sulphate.

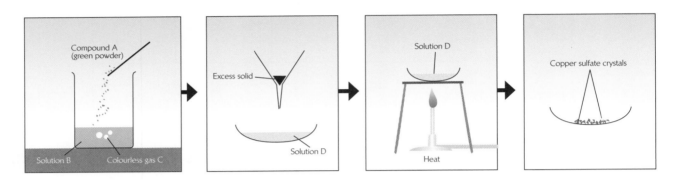

1 What does the litmus paper tell you about solution B? _____

2 Identify (name) solution B. _____

3 What is the name of gas C? _____

4 Describe in detail a test that would prove the identity of gas C.

_____

_____

_____

5 What is the name of compound A? _____

6 Fill in the gaps to complete these two equations:

   sulfuric acid + zinc carbonate → zinc sulfate + _____ + _____

   $H_2SO_4 + ZnCO_3 →$ _____

CHALLENGE

7 Give a practical reason why Sophie left solution D to evaporate overnight, instead of drying it out over a hot flame.

_____

_____

CHALLENGE

8 Write a word equation, then a balanced formula equation, for the reaction that happened in the beaker.

_____

_____

_____

_____

Your aim for tomorrow's practical in science is to make magnesium sulfate crystals.
Your teacher will provide you with: beakers, flasks, Bunsen burner, tripod, gauze, filter funnel, filter-paper and other chemicals that you can choose for yourself.

Summarise your method for producing the salt **magnesium sulfate**:

- State which chemicals you could use. (There is more than one possible answer.)
- Write your steps **in point form**.
- Describe **safety precautions** that will be needed at some steps.
- Write a word or formula **equation** for the reaction. (Bonus if you can give a second equation for a second possible reaction.)

_____

_____

_____

_____

_____

_____

_____

_____

_____

_____

_____

_____

_____

_____

_____

_____

_____

_____

_____

_____

ISBN: 9780170214667

13 Temperature experiment

Date for completion: / /

Parent sig: _____

Teacher sig: _____

This is Olivia's report on the science experiment she did together with Emma.

Aim: *The teacher told us to plan experiments to find how much a temperature difference can make to the speed of a chemical reaction.*

Method: *We set up the equipment like the drawing shows. We used the reaction of magnesium and HCl. I put exactly 10 mL HCl in the test-tube. Emma dropped in a small piece of Mg and then quickly put the stopper on the test-tube. I timed how many seconds it took before the boiling tube was half full of gas.*

Emma wrote the results in a table. We then started again and did exactly the same experiment, but with the test-tube and the acid pre-heated to 30 °C. Then we did the experiment again at 42 °C. Emma got some ice and we did one more experiment with the test-tube and acid pre-cooled to 6 °C.

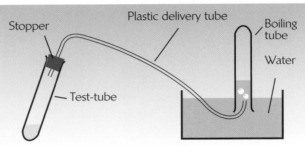

1 How many experiments (trials) are mentioned in the report? _____

2 Suggest a hypothesis they were aiming to test.

_____

_____

3 What was their independent variable? _____

4 What was their dependent variable?

_____

5 The above description mentions two factors (controlled variables) that could have affected the results, but that Emma and Olivia kept constant. Which two factors does Olivia mention?

_____

_____

6 They probably kept some other factors constant in their experiment which aren't mentioned in the report. Identify at least four of these factors (controlled variables) which they should have kept constant. For all trials they should have had:

the same _____

the same _____

the same _____

the same _____

the same _____

7 Write equations for the reactions in the test-tube.

Word equation: _____

Balanced equation: _____

8 Sketch the shape of a graph that shows the likely results of their experiments. Remember to label both axes, and put in figures for temperature.

14 Size and area

Date for completion: / / Parent sig: _____ Teacher sig: _____

SP2 Unit 3.5

Big cube
3 x 3 x 3 cm

Medium cube
2 x 2 x 2 cm

Small cube
1 x 1 x 1 cm

1 For each of these cubes, calculate its surface area and volume. Complete the table.

	3 cm cube	2 cm cube	1 cm cube
Total **surface area**, in cm ^{2}	6 x (3 x 3) = 54		
Total **volume**, in cm ^{3}	3 x 3 x 3 =		
Ratio of surface to volume	54 : ____		
Ratio simplified	____ : 1	____ : 1	____ : 1

ISBN: 9780170214667

2 From the table results, write in the missing words. Choose from this list: *bigger, smaller, same*.

Overall rule: the **smaller** an object, the _____ its surface area in relation to its

size, if they are the _____ shape. **Big** objects have a _____

surface to volume ratio, while comparatively **small** objects have a _____ surface

to volume ratio.

3 From your results in the table above, predict what are the surface to volume ratios of a cube with sides 0.5 cm long, and for a cube with sides 0.25 cm long.

0.5 cm cube _____ : 1.   0.25 cm cube _____ : 1.

4 Explain why a 1 gram lump of $CaCO_3$ will react with acid slowly, and a 1 gram lump crushed into small pieces will react much faster – if other conditions are the same in both cases.

_____

_____

_____

_____

15 Experiment plan

Date for completion: / /   Parent sig: _____
Teacher sig: _____

SP2 Unit 3.4/3.5

In an experiment to investigate surface area and reaction rate, one antacid tablet was crushed into small pieces and put into a test-tube with 10 mL of H_2SO_4 at 23°C. The rate at which bubbles of CO_2 gas came off was measured. Next, an identical tablet was used in the same way, but not crushed.

1 What was the independent variable in this experiment?

_____

2 List three controlled variables in this experiment. (Ones that were kept the same.)

_____

_____

_____

3 What was the dependent variable? (The one used to get results.)

_____

4 Suggest a practical way of getting these results. Include a diagram as part of your description.

_____

_____

_____

_____

_____

5 Describe what results you would expect to get, using your own method.

_____

6 Only two tablets were used. Explain why 10 tablets would have been better, and suggest how they could be used in this experiment.

_____

_____

_____

_____

16 Collision theory

Date for completion: / /

Parent sig: _____

Teacher sig: _____

SP2 Unit 3.4

1 A reaction can only happen when particles collide with each other. To increase the speed of reaction, the chance of colliding needs to be _____, such as by increasing concentration, the

_____, or the _____.

2 Four factors can affect the rate of reaction: **concentration**, **surface area**, **temperature**, **catalysts**. Match one of these four to each of the statements below.

Factor	Statement
	Particles move faster and bump into each other more often
	Can be changed by evaporating water from the solution
	Cut potatoes cook faster than whole potatoes
	Hydrogen peroxide breaks down faster if some MnO_2 is added
	Putting meat in the fridge slows the rate of decay
	The more particles per mL, the greater the chance of collisions
	Starch is broken down by an enzyme in your saliva
	Grinding up the solid into a powder
	Heating a solution
	More places are available for the reacting particles to bump into
	Something that speeds up the reaction, but is not itself used up
	Add more solvent to the solution
	Put some more baking powder into the cake mix
	A hot flame will fry the egg faster

3 Write what you understand by the term 'the rate of a chemical reaction'.

_____

_____

_____

ISBN: 9780170214667

17 Lost?

Date for completion: / /

Parent sig: _____

Teacher sig: _____

Chemical Science

Finn and Bevan put a lump of calcite in a beaker and placed the beaker on an electronic balance. They added 50 mL dilute HCl, and recorded the mass of the beaker and its contents at the start, then every four minutes until all the calcite had dissolved. The table shows their data. Mass has been lost from the beaker because a gas has been made, and the gas has escaped.

Time (minutes)	Total mass of CO_2 lost since the start (grams)
Start	0
4	0.14
8	0.24
12	0.32
16	0.36
20	0.38
24	0.38
28	0.38

$CaCO_3$ $CaCl_2$?
H_2O ? HCl
CO_2?

1 Plot a graph for their data, with time on the horizontal axis. (Remember the TADPL guidelines for drawing graphs.)

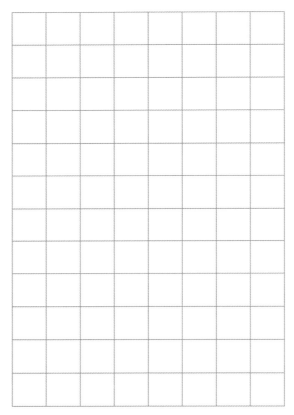

2 Calcite is a form of calcium carbonate. Write a word equation for the reaction.

_____

CHALLENGE

3 Write a chemical equation using formulae for this reaction.

_____

4 When was the reaction rate fastest? Suggest two reasons why the reaction became slower later.

_____

_____

5 Draw and label two more lines (A and B) on your graph to predict what would happen to their results if:

  A They did the same experiment with the calcite broken up into fine granules.

  B They did the same experiment with 50 mL water added to the 50 mL HCl.

18 Meet the metals

Date for completion: / /

Parent sig: _____

Teacher sig: _____

SP2 Unit 3.7

1 Complete the chart below to explain why certain metals have specific uses around the home. For the left column choose from: _gold, lead, tungsten, copper_.

Metal	Used for	One or two properties which make it useful in this kind of situation
iron/steel	car bodies	
	pipes	Unreactive with air and water
	jewellery	Malleable, unreactive, has lustre
aluminium		
	light bulb filaments	
	roof flashings	Soft, bendy, unreactive
	base of many pots and pans	
chromium		
titanium		

2 Match these words to the metal properties described in the right column: _Lustre, Malleable, Conductivity, Colour, Density, Melting point._

Word	Metal property
	Very high (except for mercury)
	Mostly silver-grey (except gold and copper)
	Tightly packed atoms
	Shiny
	Heat and electrical current flow easily
	Can be hammered into new shapes

3 Fill in the blanks to show which metals are used to make the following alloys:

Stainless steel: _____ and _____ and _____

Brass: _____ and _____

Bronze: _____ and _____

ISBN: 9780170214667

19 Metal or non-metal?

Date for completion: / /

Parent sig: _____
Teacher sig: _____

SP2 Unit 3.7/3.8

Decide whether each of these statements applies to *metals*, *non-metals* or *both*.

1 Symbols arranged on the left side of the periodic table. _____

2 Many are poisonous to people and animals. _____

3 Most elements are in this category. _____

4 At normal temperatures, some of them are gases. _____

5 They are considered to be chemical elements. _____

6 Symbols arranged mostly on right side of periodic table. _____

7 At 20 °C temperature, one of them is a liquid. _____

8 When clean, most of them are shiny (have lustre). _____

9 Most kinds are malleable and ductile. _____

10 Tend to give up electrons when they react. _____

11 Tend to take electrons when they react. _____

12 They are all good conductors of heat. _____

13 Most are good conductors of electricity. _____

14 Most are poor conductors of it electricity. _____

15 Some occur in uncombined element form in nature. _____

16 As solids, they tend to be brittle. _____

17 Some kinds burn in oxygen at high temperatures. _____

18 Some kinds react with acids to produce hydrogen. _____

19 Tend to form positive ions. _____

20 Tend to form negative ions. _____

ISBN: 9780170214667

20 Reactivity of metals

Date for completion: / /   Parent sig: _____
                              Teacher sig: _____

Chemical Science

1 Arrange these eight metals from least to most reactive by writing both their names and symbols in place above the arrow.

Calcium
Copper
Sodium
Zinc
Lithium
Magnesium
Iron
Gold

Least active ——————————————————→ Most active

2 Of the eight metals listed above, only two can be found in pure element form in nature. Which two? Why?

_____
_____
_____

3 Explain why sodium metal is kept under a layer of oil for safety reasons.

_____
_____

21 Metal properties

Date for completion: / /   Parent sig: _____
                              Teacher sig: _____

1 Explain why metals are good conductors of electricity.

_____
_____
_____

2 Explain why metals are good conductors of heat.

_____
_____
_____

3 Which group of metals in the periodic table reacts most strongly with water? The group that includes Al, Mg, or Na?

_____

4 Name three metals that do not react with dilute acids or with water.

_____

ISBN: 9780170214667

5 Explain in what way a galvanised nail is different from an ordinary nail, and in what situations galvanised nails are more useful.

_____

_____

_____

22 Word equations

Date for completion: / /

Parent sig: _____
Teacher sig: _____

SP2 Units 3.3–3.12

1 Complete word equations A–E, each of which represents general types of reaction involving metals and/ or acids.

A metal   +   oxygen   → _____

B metal   +   acid   → _____   +   _____

C metal   +   water   → _____   +   _____

D acid   +   base   → _____   +   _____

E acid   +   carbonate   → _____ + _____ + _____

2 Complete these word equations, all of which represent specific chemical reactions.

magnesium   +   oxygen   → _____

zinc   +   hydrochloric acid → _____   +   _____

lithium   +   water   → _____   +   _____

calcium   +   sulfuric acid   → _____   +   _____

aluminium   +   oxygen   → _____

sodium   +   water   → _____   +   _____

sodium hydroxide   +   hydrochloric acid → _____   +   _____

sodium hydroxide   +   sulfuric acid   → _____   +   _____

calcium carbonate +   hydrochloric acid → _____ + _____ + _____

calcium carbonate +   sulfuric acid   → _____ + _____ + _____

A small piece of sodium reacting with water, and producing hydrogen gas and sodium hydroxide.

23 Rust never sleeps

Date for completion: / /

Parent sig: _____

Teacher sig: _____

SP2 Unit 3.10

An important part of experiment design is the control principle. A control is a trial identical in every way to the main trial, except for one factor. This should be a factor you are trying to investigate. The five tube experiment at right aims to find which features speed up rusting.

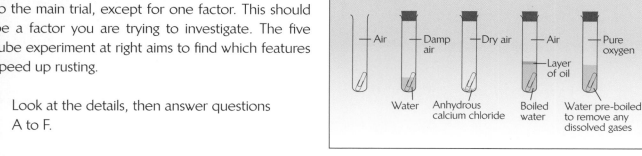

1   Look at the details, then answer questions A to F.

A   The water in Tube 4 was pre-boiled, which has the effect of removing any dissolved gases. Suggest the purpose of the layer of oil in Tube 4.

_____

B   Suggest the purpose of using anhydrous calcium chloride in Tube 3.

_____

C   Comparing which tubes tells you about the importance of oxygen? _____

D   Tube 3 is a control for which other tube(s)? _____

E   Apart from varying the amounts of water and oxygen, the five paper clips should have been treated exactly the same way (controlled variables). Suggest two or more ways in which all five paper clips and their conditions should have been the same.

_____

_____

_____

F   Suggest any practical way of estimating the amount of rust in paper clips 1 – 5, once the experiment is over. (You don't have a magic instrument for measuring rust!)

_____

_____

_____

_____

2   Two practical ways of protecting a steel trailer from rusting in a damp environment: (A) wiping it with a thin layer of oil; (B) having the steel galvanised, which adds a thin layer of zinc metal. Explain how each of these methods works.

A   Oil: _____

_____

_____

Chemical Science

CHALLENGE

Chemical Science

B Galvanising: _____

_____

_____

_____

_____

_____

_____

_____

24 Salt and rust

Date for completion: / /

Parent sig: _____

Teacher sig: _____

SP2 Unit 3.10

Max lives close to a beach. One problem of living there is that any iron objects like spades, and other garden tools, tend to rust rapidly if left outside.

One answer would be to never leave anything outside, but Mum asks Max to find another practical way of reducing rust damage. He already understands that salt makes things rust faster, and that oil will give protection. But he doesn't want everything covered in a thick layer of oil. Max gets organised, and plans a series of experiments to find out how little oil is needed to give rust-protection. All he has is a bag of kitchen salt, 12 plastic cups, plenty of new non-galvanised nails, some oil, and patience.

Plan some experiments for Max, with the aim of finding the least amount of oil needed in different salt conditions. Describe your method in a series of numbered points.

CHALLENGE

_____

_____

_____

_____

_____

_____

_____

_____

_____

_____

_____

_____

_____

_____

_____

ISBN: 9780170214667

Making electricity

Date for completion: / /

Parent sig: _____
Teacher sig: _____

All metals tend to give away electrons. A chemically-active metal like Mg is more willing to give away electrons, compared to a 'lazy' metal like Cu. This diagram shows a simple electrochemical cell. The electrolyte is a dilute HCl solution.

This table gives the voltages of 10 metals when each is paired with hydrogen. (Slightly different results to those from testing the reactivity of metals with acids.)

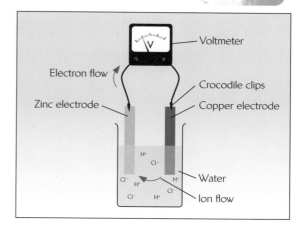

Metal	Li	Na	Al	Zn	Fe	Pb	Cu	Ag	Au
Voltage	-3.03	-2.71	-1.70	-0.76	-0.44	-0.13	+0.34	+0.80	+1.50

When answering questions 1 – 4, in each case calculate voltage from the **total difference** between the two elements named.

1 Calculate the theoretical voltage from a zinc-lead cell.

_____

2 Calculate the theoretical voltage from a sodium-lead cell.

_____

3 Calculate the theoretical voltage from a sodium-copper cell.

_____

4 Of the metals listed, predict which pairing would give the highest voltage. Calculate what that voltage would be.

_____

5 Suggest why cells with these metals are not for sale.

_____

6 Explain the purpose of an electrolyte (the solution) in the cell drawn above.

_____

_____

7 What is the main chemical fact that causes electricity to flow in simple cells?

_____

_____

_____

Chemical Science

CHALLENGE

ISBN: 9780170214667

Chemical Science

1 Rocks seldom or never contain useful metals like iron, lead or zinc in element form. They are mined as compounds. Give the names of the following metal compounds:

Fe_2O_3 _____

PbS _____

ZnO _____

Al_2O_3 _____

$CuCO_3$ _____

2 Explain why gold ore contains pure gold, and not a gold compound.

_____

_____

3 Explain what is meant by an 'ore'.

_____

_____

4 For most metals, after the ore has been mined the metal has to be chemically extracted from its compounds. This process is known as **reduction**. In many situations, reduction involves using carbon as an oxygen-acceptor.

metal oxide + carbon + energy → metal + carbon dioxide

Complete and balance these two chemical equations showing reduction of metal compounds using carbon.

ZnO + _____ → _____ + _____

Fe_3O_4 + _____ → _____ + _____

A lump of iron ore, mostly the compound iron oxide.

Stainless steel is an alloy: about 80% iron, 20% nickel and chromium.

ISBN: 9780170214667

Chemical Science

Coltan is the industrial name of a metallic ore named 'columbite-tantalite'. Two elements are extracted from coltan: the rare metals niobium and tantalum. Tantalum is used in capacitors for high performance electronics that need to be reliable and compact. It is used in the production of consumer electronics including cell phones, DVD players, computers, and video-game consoles.

Worldwide production of coltan in 2009 was 650 tonnes, equivalent to only a few dozen truck-loads. About 12% was mined in Australia, 27% in Brazil, 4% in Canada, and 57% in several countries in Africa – the Democratic Republic of the Congo (DRC) is a major producer.

This means that the Internet-enabled smart phone in your pocket, or your handheld games console, very likely contains little pieces of eastern Congo.

Eastern Congo has been the site of savage and ongoing civil wars for many years, with private armies struggling for control of the area. Some of the conflict is tribal. An estimated 6.9 million people have been killed there since 1998. Much of the money for weapons and for paying soldiers comes from the sale of coltan and other minerals. Much of the mining is done by local people and their children, who have few other sources of income. Mining methods are highly polluting. The DRC central government has little influence in the war zones.

Export of coltan from the eastern DRC to European, American and Asian markets has been cited by experts as helping to finance the present-day conflict in the Congo. They claim that much of the finance sustaining the civil wars in Africa, especially in the DRC, is directly connected to coltan profits. Likened to blood diamonds, some claim that coltan mining is causing ecological damage and human rights abuses.

1 Underline key words in the above information.
2 Write a summary, in the form of two bullet points for each of the above five paragraphs.

_____

_____

_____

_____

_____

_____

_____

_____

_____

_____

_____

ISBN: 9780170214667

Chemical Science

ACROSS

3 The scientific name for rust
5 Any property that involves reactions
7 Able to be bent or hammered into new shapes
9 Most reactive of the first 20 elements

DOWN

1 Able to be stretched into wires
2 Gas produced when a metal and acid combine
4 The removal of oxygen from a metal compound
6 Heated metal that is in a liquid state
7 An exception to the high melting point property of metals
8 A mixture of different metal elements

Like most other metals, copper is shiny, a good conductor of electricity, and also ductile – easily made into flexible wires.

ISBN: 9780170214667

Team carbon

SP2 Unit 3.13

With thousands of different carbon compounds, scientists have organised them into 'families' that share similar features. Three examples of families are:

- **hydrocarbons** contain only the elements hydrogen and carbon
- **alcohols** are of similar shape to hydrocarbons, but always have an –OH group
- **carbohydrates** always have hydrogen and oxygen in the ratio 2:1.

The first part of a carbon compound's name tells you how long the carbon chain is. This table lists the first eight.

Number of carbon atoms	1	2	3	4	5	6	7	8
Compound name starts with	meth-	eth-	prop-	but-	pent-	hex-	hept-	oct-

The second part of a name tells you which family it belongs in. Examples:
- the names of simpler **hydrocarbons** end in **-ane**
- the names of **alcohols** end in **-anol**

We draw molecule shapes by using letters to represent atoms, and straight lines to represent bonds. C always has four bonds, H always has one bond, O always has two bonds. Examples:

1 For the following eight molecules, make drawings similar to those above.

ethane

ethanol

propane

propanol

butane

butanol

ISBN: 9780170214667

hexane (straight-chain)

octanol (straight-chain)

2 Draw three different possible atom arrangements for the compound pentane, which is C_5H_{12}. Different arrangements like this are known as **isomers**.

---

30 Plastics

| Date for completion: | / / | Parent sig: _____ |
| | | Teacher sig: _____ |

SP2 Unit 3.14

1 For each of the following four objects, suggest what material(s) were used in its place before plastics were invented. For each object, suggest at least one advantage of using plastic.

Plastic shopping bag _____

_____

_____

Plastic milk bottle _____

_____

_____

Car bumper _____

_____

_____

Water pipes_____

_____

_____

2 Chemically, all plastics are polymers. This diagram shows how one kind of plastic is made. When ethene is compressed and heated, these monomers link together to form the plastic we know as poly-ethene, or polythene for short. (In Greek: *poly* = many, *mono* = one, *meros* = part.)
In your own words, explain the difference between a monomer and a polymer.

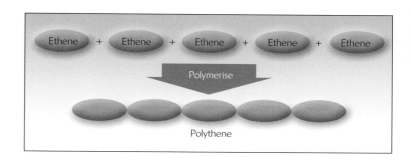

Ethene + Ethene + Ethene + Ethene + Ethene

Polymerise

Polythene

_____

_____

_____

3 Suggest the monomer name of each of these plastics:

polystyrene _____

polyvinyl chloride (PVC) _____

polyurethane _____

polypropylene _____

31 Carbohydrates and lipids

Date for completion: / /

Parent sig: _____
Teacher sig: _____

SP2 Unit 3.15

Complete the following sentences by choosing from this list. Each word is used once: *orange, plants, polysaccharides, Benedict's, black, carbon, starch, glucose, monosaccharides, iodine, fats, animals, fructose, glycerol, fatty acid, oils, disaccharides, sucrose, blue, cellulose.*

Carbohydrates are a major group of organic substances. All of them contain the element

_____(1), with H and O in the ratio 2:1, just as in water. Most carbohydrates

are built up of units of 6-carbon simple sugars, also known as _____(2).

Examples: _____(3) and _____(4), which are common

in many plant foods. The chemical word for double sugars is _____(5).

'Kitchen sugar', a.k.a. _____(6), is a well-known example. Multi-sugars,

a.k.a. _____(7) can have up to thousands of glucose molecules joined

together. Well-known examples of this type are _____(8), the main part

of wood and paper and cotton; and _____(9), commonly found in grain

foods. An easy way of finding out if there are monosaccharides in a solution is to heat it with

_____(10) solution. If the solution turns _____(11)

this shows that sugars are present; if it remains a clear _____(12)

ISBN: 9780170214667

colour that will show that sugars are absent. The usual test for starch is to add some

_____(13) solution without boiling: a _____ (14)

colour is a positive result.

'Lipid' is a collective name for fats and oils. We use the word _____

(15) for lipids that are solid at 20 °C, and _____(16) for those that

are liquid. Mostly, fats are derived from _____(17) and oils from

_____(18). Although there are many kinds of lipid, in all cases when they

are digested they break into three molecules of _____ _____ (19), and one

molecule of _____(20).

All carbohydrates are made up of small sub-units called sugars.
The scientific name for sugar is saccharide.

Single sugars (monosaccharides) eg. glucose and fructose

Double sugars (disaccharides) eg. sucrose, maltose and lactose

Polysaccharides eg. starch, glycogen and cellulose

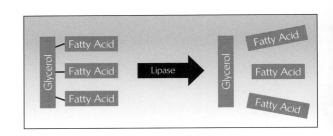

32 Detergents and soaps

Date for completion: / /

Parent sig: _____

Teacher sig: _____

SP2 Unit 3.16

1   The following sentences have had their three main components jumbled. Rewrite each sentence correctly.

Subject	Verb phrase	Object
Soaps	is an example of	the water-attracting end of a molecule.
Synthetic detergents	refers to	oil droplets in water.
The word hydrophilic	are made from	crude oil derivatives.
The word hydrophobic	are made from	a strongly polar non-organic molecule.
Water	is a suspension of	plant oils or animal fats.
An emulsion	refers to	the water-avoiding end of a molecule.

Soaps _____

_____

Synthetic detergents _____

_____

The word hydrophilic _____

_____

The word hydrophobic _____

_____

Water _____

_____

An emulsion _____

_____

2 The diagrams below show soap removing grease from a (green) surface. Under each of the diagrams, write in your own words a caption explaining why the soap molecules are behaving as they are, and describing what their eventual effect will be on the grease.

_____   _____

_____   _____

_____   _____

_____   _____

_____   _____

_____   _____

33 Drugs

Date for completion: / /

Parent sig: _____

Teacher sig: _____

SP2 Units 3.17/2.28

The word 'drugs' can mean medically-useful substances (like antibiotics and painkillers), or it can mean mind-altering substances. Some chemicals are in both categories, for example morphine.

1 Classify each of the mind-altering drugs **A** to **G** as having one or two of the following effects: *painkiller, anaesthetic, stimulant, depressant.*

A morphine _____

B lignocaine _____

C ecstasy _____

D nicotine _____

E nitrous oxide _____

F heroin _____

G alcohol _____

ISBN: 9780170214667

2 From the above list, identify which drug is:

A used as a painkiller (local anaesthetic) by dentists _____

B used as a general anaesthetic during operations _____

C derived from opium poppies _____

3 The drawing shows a molecule of morphine. Count the number of atoms of:

A carbon _____

B hydrogen _____

C oxygen _____

D nitrogen _____

4 Morphine is a very useful substance, so explain why its availability is strictly controlled by law.

_____

_____

5 Below are the names of two drugs banned in sport. For each, describe its effect on the body, and what sport(s) it is sometimes used in, and how it could give temporary advantage to a user.

A erythropoietin (EPO)

_____

_____

_____

_____

B beta blockers

_____

_____

_____

_____

6 Alcohol and tobacco can affect sports performance, so explain why they are not banned in sport.

_____

_____

_____

_____

For this activity, refer to the periodic table (page 160). Numbers at the top of each column apply to a group of similar elements. Group numbers also help you work out how many outer electrons each atom has.

1   Classify each of the following elements as an **alkali metal** (group 1), an **alkaline-earth metal** (group 2), **transition metal** (groups 3-12), **metalloid** (green boxes), **halogen** (group 17), or **noble/inert gas** (group 18).

Chlorine   _____

Silver   _____

Argon   _____

Sodium   _____

2   How many **outer** ('valence') electrons do each of these elements have?

•   magnesium _____

•   oxygen _____

•   iodine _____

•   lithium _____

3   Which of the following ions could exist? If the given ion can't possibly exist, write that element's correct ion.

• N^{1+} _____

• F^{1-} _____

• P^{6-} _____

• He^{3+} _____

• Al^{2+} _____

• Mg^{2+} _____

4   Explain why fluorine is a reactive element, but argon is not.

_____

_____

5   Explain why calcium loses two electrons when it forms ions, but chlorine gains an electron.

_____

_____

_____

6   Explain why sodium and potassium have similar reactivity. Include the words 'outer/valence electrons' in your answer.

_____

_____

_____

_____

ISBN: 9780170214667

Physical Science

1 Introducing calculations

SP2 Unit 4.1/4.2

Date for completion: / /
Parent sig: _____
Teacher sig: _____

Many questions in science involve calculations that begin with a maths formula.

> Here are some formulae you will be dealing with:
>
> $v = d / t$ $a = \Delta v / t$ $w = f d$
> $F = m a$ $P = w / t$ $P = V I$
> $v = d / t$ is the main formula at this stage.
>
> It's a good idea to set out this type of calculation in three steps: F S C.
>
> **F**ormula: choose the correct formula, and write it down.
> **S**ubstitute: put numbers in the place of letters.
> **C**alculate: then round your answer to 3 S.F. (significant figures). Use the correct units.
>
> You sometimes need to rearrange an equation at the start.
> The following method works for any equation that has three parts.

For speed calculations it is best to use **v** for velocity and not **s** for speed. Reason: **s** is the symbol used for seconds.

1 The direct distance from Auckland to Sydney is 2150 km. Flight time is 2.5 hours. Calculate the average speed.

formula $v = d / t$

substitute   v = 2,150 km / 2.5 hours

calculate   _____

2 A family drives 1,099 km from Wellington to Cape Reinga in 14 hours, excluding stops. Calculate the average speed for the journey.

formula   _____

substitute   _____

calculate   _____

3 Your thumbnail grows 6 cm in one year. Calculate its speed in mm per week (to three S.F.).

formula   _____

substitute   _____

calculate   _____

4 Caleb's 1-hour training run average speed was 2.5 ms^{-1}. How far did he run? (In kilometres.)

formula   d = v t

substitute   _____

calculate   _____

5 The drive from Nelson to Timaru is 587 km. Nicole averages 80 km/hr, including stops. How long did her trip take? (In hours.)

formula   _____

substitute   _____

calculate   _____

6 A cheetah has a top speed of 25 ms^{-1}, but can only sprint for 8 seconds. What distance can it cover in this time?

formula   _____

substitute   _____

calculate   _____

2 More speed

Date for completion: / /

Parent sig: _____

Teacher sig: _____

SP2 Unit 4.2

1 Usain Bolt won the 2008 Beijing Olympics 100 metres event in a time of 9.69 s. (A) Calculate his average speed. (B) Explain why his top speed was faster than his average speed.

_____

_____

_____

_____

2 On a road trip, Holly's family covered 90 km in the first hour, then stopped for three quarters of an hour, then drove an average of 95 km/hr for half an hour, then stopped again for 30 minutes, then continued at an average of 92 km/hr for the next two hours. Calculate the total distance covered.

ISBN: 9780170214667

3   A runner has a stride (step) length of 1.8 metres, and takes 2.4 strides each second. (A) What is his speed? (B) 1 ms^{-1} = 3.6 km/hr, so what is his speed in km/hr? (C) How long would he take to cover half a kilometre at this speed?

_____

_____

_____

_____

_____

_____

4   A skydiver is falling at a maximum speed of 35 ms^{-1}. She reached this speed at 2,500 metres, and opened her parachute at 500 metres above the ground. Calculate the time spent in freefall at 35 ms^{-1}.

_____

_____

_____

_____

_____

_____

_____

> Note: you can write speed (velocity) units as m/s or ms^{-1}. At this stage it is not important which, but physics textbooks generally use ms^{-1}.

---

3 ● Distance-time graphs

| Date for completion: | / / | Parent sig: _____ |
| | | Teacher sig: _____ |

These graphs show distances travelled in different situations, and measured in different ways.

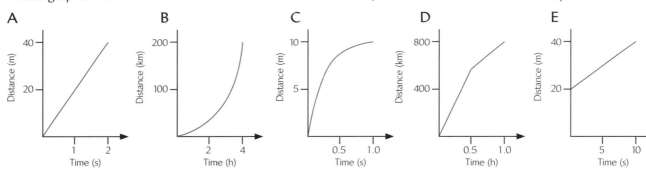

1   Describe in plain words the movement represented by each of these graphs. Example (not shown above): 'Not moving at the start, then a steady speed for 10 seconds, then accelerated for 5 seconds until 200 m from the start.'

Graph A   _____

_____

Graph B   _____

_____

Graph C   _____

_____

Graph D   _____

_____

Graph E   _____

_____

2 For graph E, calculate the average speed. Remember to write the units!

formula: _____

substitute: _____

calculate : _____

Suzy's dog

Date for completion: / /

Parent sig: _____
Teacher sig: _____

SP2 Unit 4.1/4.3

Suzy takes her dog for a brief walk and run. The table gives her distance from the front gate at different times.

Time (s)	Distance (m)
0	0
10	10
20	20
40	30
60	95
80	175
100	180
120	180
200	180
320	80
340	60
380	30
400	15
420	7
440	2
480	0

ISBN: 9780170214667

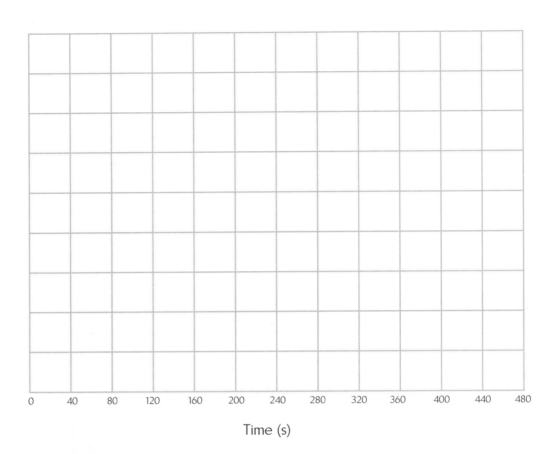

Distance from home (m)

Time (s)

1   Draw a line graph using the numbers in the table. Remember the 'TADPL' guidelines. Notice that the time intervals in the table are not even, but the graph should have even intervals on both the horizontal and the vertical axes.

Use the graph to answer the following:

2   How far from home did Suzy get? _____

3   How many minutes was the walk, in total? _____

4   What was the total distance of her 'journey'? _____

5   Use these words to clearly label different parts of your graph: stopped, constant speed, acceleration (speeding up), deceleration (slowing down).

6   During which time interval was Suzy moving quickest? _____.
    Without doing calculations, how can you judge she was quickest at this time?

    _____

7   Calculate her speed at the quickest 40 second interval.

    formula: _____

    substitute: _____

    calculate: _____

8   If 1 metre per second = 3.6 kilometres per hour, calculate Suzy's maximum speed in km hr^{-1} .

    _____

Date for completion: / /

Parent sig: _____
Teacher sig: _____

SP2 Unit 4.4

CHALLENGE

This task can also be used for Level 1 science.

> Definition: Acceleration is the rate of change in speed or velocity.
>
> Formula:
> $$\text{acceleration} = \frac{\text{velocity change}}{\text{time taken}}$$
>
> also written as: $a = \Delta v / t$
>
> The units for acceleration may seem strange at first:
> 'metres per second, per second',   m/s/s,   more usually written as ms^{-2}

1 From a standing start, a fully laden Boeing 767 reaches lift-off speed of 90 ms^{-1} in 45 seconds. Calculate its acceleration down the runway. (Remember to use the correct units.)

   formula: _____

   substitute: _____

   calculate: _____

2 A sprinter gets from zero to her top speed of 8 ms^{-1} (28.8 km/h) in 4.5 s. Calculate her acceleration.

   formula: _____

   substitute: _____

   calculate: _____

3 To avoid a collision, Adam brakes hard, slowing from 20 ms^{-1} (72 km/h) to zero in 4 seconds. Calculate his deceleration.

   formula: _____

   substitute: _____

   calculate: _____

4 After cornering, an F1 racing car accelerates from 30 ms^{-1} (108 km/hr) to 65 ms^{-1} (234 km/hr) in 5 seconds. Calculate its average acceleration in ms^{-2} .

   formula: _____

   substitute: _____

   calculate: _____

An object in freefall in Earth's gravity accelerates downwards at a speed of 9.8 ms^{-2}.
This acceleration is known as 'g', and is used in many physics calculations.
So '2g' means an acceleration (or deceleration) of 19.6 ms^{-2}.
Sometimes the figure of 9.8 ms^{-2} is rounded to 10.
Actual freefall acceleration may be less than 9.8, due to air resistance.

Physical Science

ISBN: 9780170214667

Physical Science

1 For each of these speed/time graphs, describe the types of motion using words like: *acceleration, deceleration, constant speed, lower acceleration.*

A

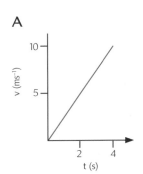

B

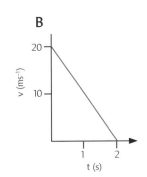

C

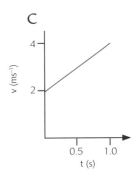

_____   _____   _____

2 For each of these speed/time graphs, describe changes in the types of motion, and when the changes occured. Use words like: *acceleration, deceleration, constant speed.*

A

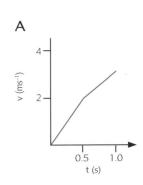

B

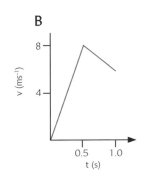

C

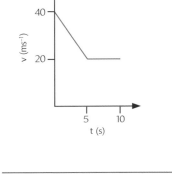

_____   _____   _____

_____   _____   _____

_____   _____   _____

3 For the middle graph above (2 B), calculate:
Acceleration over the first 0.5 seconds

formula: _____

substitute: _____

calculate: _____

Acceleration from 0.5 to 1.0 seconds

formula: _____

substitute: _____

calculate: _____

Ticker tape

Date for completion: / /
Parent sig: _____
Teacher sig: _____

SP2 Unit 4.4

This is a full-size copy of a strip of ticker tape produced when Kelly and Sam measured movements of a trolley in science class. The dots are exactly 1/50 of a second apart. The tape went through the ticker-timer from right to left, so the first dot is at the left end.

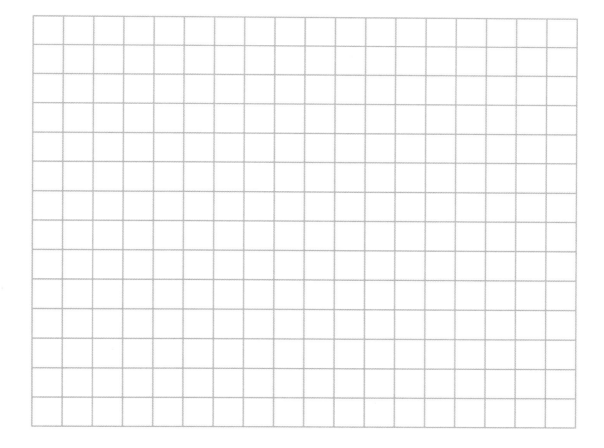

Start

Finish

1   Use a ruler to carefully measure the distance between adjacent dots, to the nearest millimetre. Start with measuring the space next to dot number 1 on the left. Record the results in this table.

Dot	1	2	3	4	5	6	7	8	9	10	11	12	13	14	15	16	17
Distance from start (mm)																	

2   Use the results in your table to draw a line graph. Put dot numbers 1-17 on the horizontal axis, and label this axis 'Time, 1/50 s'. Label the vertical axis 'Distance from start, mm.' Neatly place the 17 dots and join them. Give the graph a title.

3   On your graph, write at least two descriptions of the trolley speed, such as 'steady speed here' and 'fastest acceleration here.'

ISBN: 9780170214667

1 Complete the sentences below using the words provided in this list. Some may be used twice:
speed, pull, direction, push, shape, twist.

A force is defined as a _____ or a _____

or a _____. When you open your front door from the inside,

you _____ the door handle. For flexible materials, a force can

also make things change _____. If forces are not balanced

they can make something change _____ and also perhaps

_____.

2 The diagrams below show some forces in action. The arrows next to each show the **direction** that the force is acting.
- State whether the force's arrow represents a **push or a pull**.
- State **which** specific object the force is acting on.
- Describe the likely **effect** the force has on that object.

_____
_____
_____
_____

_____
_____
_____

_____
_____
_____
_____

_____
_____
_____

Physical Science

Force pairs

Date for completion: / /

Parent sig: _____

Teacher sig: _____

Newton's Third Law of Motion can be summed up as:

'Every force causes an equal and opposite force.'

This is also known as 'action and reaction', even when there is no movement. For stationary objects this is easy to understand. If your weight pushes down on the floor with a force of 700 N, then the floor 'reacts' by pushing upwards on your feet with a force of 700 N.

With a moving object, it helps to identify the separate forces acting on it. Then it becomes easier to see whether it is changing speed, or moving at a constant speed.

1 Force A is 600 N. How big is force B?

_____

2 Describe what is causing force A.

_____

3 Explain in detail what is causing force B.

_____

_____

_____

_____

_____

4 When the diver takes a small step backwards off the board, identify the main force acting on her.

_____

5 For the next second of freefall during the dive, give one word to describe the 'type' of movement this force will have on her. (Apart from the word 'down!) _____

6 As she falls, a comparatively small 'external' force begins to affect her dive. What is this other force?

_____

7 The diver's mass is 61.2 kg. Use the formula **F = ma** to calculate her downward acceleration as she falls. (See Unit 14: 'It's the law'.)

_____

_____

_____

_____

8 Identify and describe forces N, O and P.

M: pull of dog on chain

N: pull of _____ on _____

O: pull of _____ on _____

P: pull of _____ on _____

Physical Science

Newton's Third Law of Motion says that each force causes an equal and opposite force at the point of contact.

However, not all forces are equal. If the tree branch breaks, there will suddenly be no upward force. The main force pulling on the man will be gravity. Result: he will accelerate downward.

For each of the diagrams below:
- Draw force arrows of the main forces pushing or pulling on the blue figure.
- Put a short description next to each arrow.
  Example: Parachute cords pulling on blue figure.

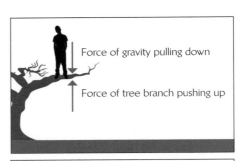

Force of gravity pulling down

Force of tree branch pushing up

Force of gravity pulling down

Physical Science

1 The arrows next to each of these cars show the directions and comparative sizes of four different forces, but do not show the exact points each force is pushing or pulling against,
   • Label each force arrow as: **weight**, **support**, **friction**, or **wheel push.** (It is technically more correct to say that the engine turns the wheels, and that the road surface pushes back on the wheels.)
   • State whether the car is **slowing** down, **accelerating**, **at rest** (stopped), or travelling at **constant speed**.

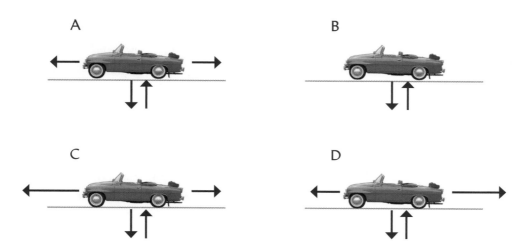

2 The gravitational force exerted on each kilogram of matter at the Earth's surface is about 10 newtons (to be more exact, 9.8 N kg⁻¹).

   A What is the weight (in newtons) of a car with mass 1,200 kg? _____

   B Ruby's bathroom scale registers 52 kg. What is her weight in newtons? _____

   C A book has a mass of 200 g. What is its weight in newtons? _____

3 Complete the sentences below using these words (use each word only once): *smooth, concrete, lubricants, more, heat, friction, brakes, rough, ice, rubs, liquid, glass.*

   You can drag an object more easily across a _____ surface than a _____ surface. This is because of _____, which happens when one surface _____ against another. Rough surfaces like _____ produce _____ friction than smooth surfaces like _____. Friction always generates _____ energy from kinetic energy. In situations like car engines and wheel bearings, designers need to reduce friction. To do this we use oil and grease, which are examples of _____. Oil works by creating a thin layer of lower-friction _____ between two solid surfaces. In many situations we depend on friction, for example when we use car _____ to slow down. Walking would be impossible without some friction, which is why it is so difficult to walk on _____.

ISBN: 9780170214667

Physical Science

Date for completion: / /

Parent sig: _____
Teacher sig: _____

1   On each of the following diagrams, clearly mark the turning point (a.k.a. pivot or fulcrum) with a small neat circle. Add and label two arrows, one labelled 'load force', and the other labelled 'effort force'. Position these two arrows at the centre of effort and load (resistance).

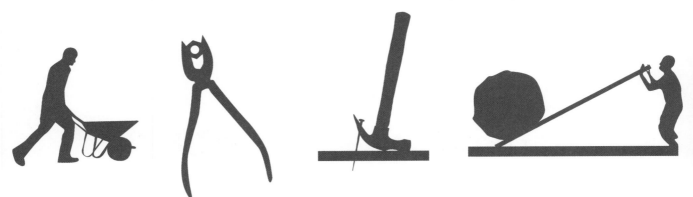

In each of the diagrams **2** to **4**, the stick is exactly balanced by the weights. All centimetre measurements are distances from the central pivot. Each blue weight is 20 N. For each drawing, estimate or calculate the different size of the weight(s) on the other side. Write your answer next to each round weight. In diagram **4**, the two weights are the same.

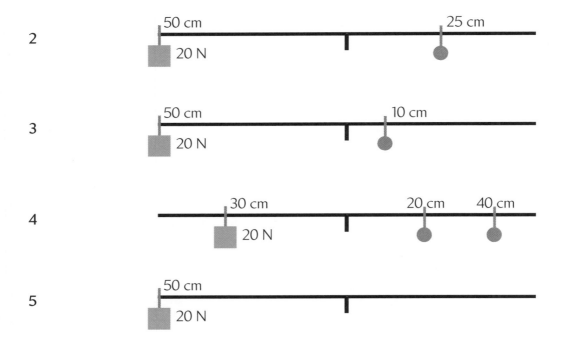

For question **5**, draw where you would place a 50 N weight to achieve balance. Give its exact position in centimetres.

Date for completion: / /

Parent sig: _____

Teacher sig: _____

SP2 Unit 4.5–4.8

1 For each of the following force diagrams, draw an arrow to show the direction of the net force, and write on the arrow the size of the net force. Example: in the first diagram the net force is 30 – 50, which is a 20 N pull to the right.

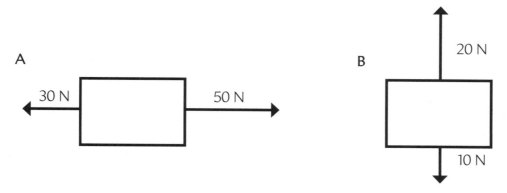

2 In situations A, B and C, a car is moving to the left along a straight road. In each case the horizontal forces acting on the car are described. Up-down forces are not shown.
   • On each car draw arrows to represent the horizontal forces. State force size.
   • Underneath, state net force size and direction.
   • Say what the motion of the car will be. Choose from: *acceleration, deceleration, stationary, constant speed.*

A The forward force is 5000 N and the frictional forces total 500 N.

The size of the unbalanced force is _____. The direction in which it acts is

_____, so the motion of the car is _____.

B The forward force is 400 N and the frictional forces total 500 N.

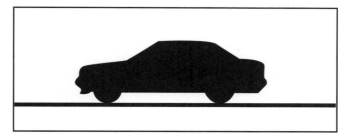

The size of the unbalanced force is _____. The direction in which it acts is

_____, so the motion of the car is _____.

ISBN: 9780170214667

Physical Science

C The forward force is 500 N and the frictional forces are 500 N.

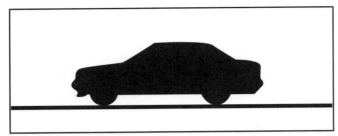

The size of the net force is _____. The direction in which it acts is

_____, so the motion of the car is _____.

---

14 It's the law

Date for completion: / /
Parent sig: _____
Teacher sig: _____

SP2 Unit 4.8

Newton's Second Law of Motion is summed up in a mathematical formula.

$$F = m\,a$$

Mass units are kg. **Acceleration** units are ms^{-2}. **Force** units are newtons, symbol N.

A one newton force gives a mass of 1 kg an acceleration of 1 ms^{-2}.
One N is not a large force. For example, a cricket ball weighs about 1.6 N.

1 Toby's car gets from zero to 20 ms^{-1} in 4 seconds. Calculate its acceleration.

formula: _____

substitute: _____

calculate: _____

The car and Toby together have a mass of 800 kg. Calculate the force needed to give this acceleration.

formula: _____

substitute: _____

calculate: _____

Note: the actual force is greater than your calculated answer because your calculated answer does not include air resistance and other friction forces.

2 Complete these sentences.

Acceleration is in proportion to the amount of force; and acceleration is inversely in

_____ to _____. These facts are known as Newton's

Second law of _____, and apply to all moving things. For example, if you

double the mass that a car engine has to pull, the acceleration will be _____

as much, assuming that _____ is the same.

Physical Science

CHALLENGE

3 From a standing start, a 747 aircraft accelerates down the runway, reaching lift-off speed of 100 ms^{-1} in 40 seconds. Calculate its acceleration.

formula: _____

substitute: _____

calculate: _____

Its total mass is 300 tonnes (300,000 kg). Calculate the force from the engines during takeoff.

formula: _____

substitute: _____

calculate: _____

CHALLENGE

4 Engineering fact: maximum thrust force from jet engines is around 220,000 N each. Calculate the proportion of maximum thrust from all four engines used in takeoff.

_____

15 Power

Date for completion: / /

Parent sig: _____

Teacher sig: _____

SP2 Unit 4.9

In science and technology, power is defined: **the rate at which work is done.** 'Work' means energy being changed from one form to another. 'Power' refers to energy conversion in all sorts of situations: muscles, machines, ocean waves, electrical appliances.

Energy is measured in joules, symbol J. Power is measured in watts, symbol W. One watt is one joule per second.

A hairdryer does work by converting electrical energy to heat + sound + kinetic energy. A 500 W hairdryer converts energy at a rate of 500 joules every second.

The formula for calculating power is:

$$\text{power (watts)} = \frac{\text{work done (J)}}{\text{time taken (s)}}$$

$$\text{power} = \frac{\text{work}}{t}$$

1 An electric forklift raises a 100 kg load up to the top shelf, a vertical height of 5 metres. This involves about 5,000 J work.

A If the 5 metre lift takes 4 seconds, calculate the power of the lift motor.

formula: _____

substitute: _____

calculate: _____

B Convert your answer to kilowatts (kW). _____

C Write an energy equation to represent the main energy conversion.

_____

ISBN: 9780170214667

2 You drag a small dinghy along the beach for 25 metres with a steady 200 N force.

A Calculate how much work you have done.

formula: work (J) = force (N) x distance (m)

substitute: _____

calculate: _____

B If the pull takes one minute, calculate your power during the task.

formula: _____

substitute: _____

calculate: _____

3 Amber, mass 50 kg, runs upstairs in 10 seconds, a vertical height of 4 metres. Calculate her power during the 'climb'.

_____

_____

_____

_____

_____

16 Missile experiment

Date for completion: / /

Parent sig: _____

Teacher sig: _____

In a class experiment, Tess and Jordan noticed that how far the tennis ball 'missile' went seemed to depend on the firing angle, as well as how much the elastic was stretched. So they did further experiments to test the angle question. Read their method before making your own planning suggestions.

Aim: To answer the question 'How much difference does firing angle make to the missile?'

Method:
- Jordan and Tess fastened a 40 cm double length of strong elastic to the side of a box as the diagram shows. A plastic cup was used to hold the tennis ball.
- They placed the whole setup at one end of a flat grassy area, and used a big protractor to measure angles.
- Tess pulled back the elastic to almost full stretch at the position marked in red, then let go, firing the tennis ball in the direction of the red arrow.
- Each time Jordan marked the spot where the tennis ball landed, then measured the distance from starting point X. They did five tennis ball firings at the red angle.
- They then repeated the 'missile' firing with the elastic at different angles: green, blue and others. They did five trials at each angle.

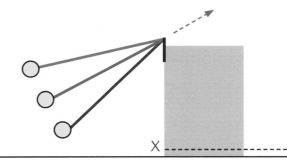

Results: for the red angle only.

Trial number	1	2	3	4	5
Distance to landing	16.2 m	16.9 m	15.9 m	6.1 m	16.6 m

A They made two kinds of measurements: the firing angle, and the distance to landing point. In their experiment, which was the **independent variable**? _____

B Which was the **dependent variable**? _____

C Suggest at least 5 **controlled variables**: factors which Tess and Jordan should have kept the same to make sure that they did fair trials when comparing angles.

the same tennis ball_____

the same _____

the same _____

the same _____

the same _____

the same _____

the same _____

D Which red result do you think should be ignored? Explain why.

_____

_____

E If you ignore this result, calculate the red average.

F Jordan and Tess eventually did trials at seven different angles: 0° (horizontal), 20°, 30°, 45°, 60°, 70°, 90° (vertical). Predict their overall results by putting in seven crosses and a curved line on this graph.

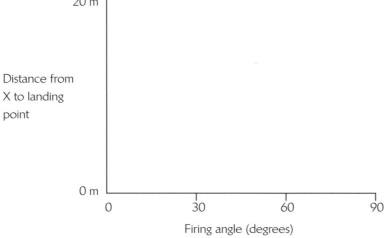

ISBN: 9780170214667

17 Conductor or not?

Date for completion: / /   Parent sig: ______ Teacher sig: ______

SP2 Unit 4.11

1 The red box contains the names of 14 different objects and substances. Which will conduct electricity? Write each in its correct column.

steel scissors; cork; glass bottle; candlewax; wooden block; paper; seawater; key; pure water; string; plastic comb; 20 cent coin; pencil graphite; Al foil.

Conductors	Insulators

2 Explain why metals are good conductors of electricity, but most non-metals are not.

______

______

______

3 Draw a simple circuit that you could use to test whether or not a piece of charcoal (carbon) is a good conductor of electricity.

Physical Science

ISBN: 9780170214667

1 Complete the following sentences, choosing from among these words (not all words will be used): *gain; static; lightning; electrons; loss; spark; charge; friction; negatively; positively; protons.*

Static electricity can occur with _____ (1), when two insulating materials are

rubbed together. Some _____ (2) get rubbed off one material and onto

another. This creates a difference in _____ (3). One material now has

surplus electrons, which means it has become _____ (4) charged.

The other material has had an electron _____ (5), which makes it

_____ (6) charged. If the charge cannot quickly flow back again, the

charge is said to be _____ (7). If a big enough charge builds up, a

_____ (8) can jump from one object to another. Static electricity can be

dangerous. For example, charges build-up in a cloud can cause _____(9).

2 The drawing shows a balloon stuck to a wall by the attraction of opposite charges. The balloon was made negative by rubbing it on clothing. Explain how the wall became positive.

_____

_____

_____

_____

_____

Physical Science

ISBN: 9780170214667

Complete this table listing electrical components, their symbols, and their function.

Physical Science

	Component name	Symbol	Function
1			
2			Converts chemical energy to electrical energy; several cells in series.
3	Connecting wire		
4			
5			Converts electrical energy to light and heat.
6			Converts electrical energy to heat. Opposes current flow.
7			Measures current (flow of charge).
8			Measures the energy difference (per coulomb of charge) across a component.
9			

Date for completion: / / Parent sig: _____
Teacher sig: _____

1 Write out a list of all the equipment you would need to set up circuits A and B.

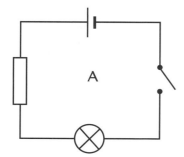
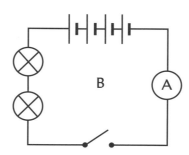

A _____

_____

B _____

_____

2 Draw circuit diagrams for circuits A and B.

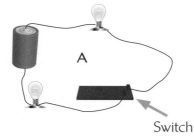

Switch

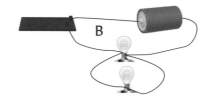

3 Draw a diagram for a circuit with one cell and two lamps (bulbs) in series.

Physical Science

ISBN: 9780170214667

4   Draw a circuit diagram with two lamps in parallel, one lamp in series, and one cell.

5   Draw two circuit diagrams:
    A   two cells and a lamp in series
    B   two cells in parallel and two lamps in series.

21   Electron exercise

Date for completion: /   /   Parent sig: _____
Teacher sig: _____

SP2 Unit 4.14

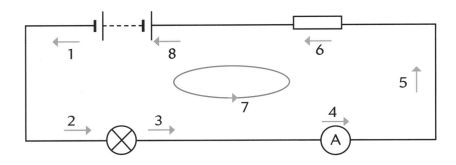

Note: this diagram shows the direction of electron flow, − to +. Normally we show 'conventional current' flowing + to −.

Complete the sentences below with a word to fill each gap. Choose from List A for the first gap, and from List B for the second gap. You may use words more than once. The numbers refer to numbered parts of the circuit. The blue arrows show the direction of electron flow.

List A: *across, along, around, away from, through, out of, into, towards*
List B: *battery, lamp, circuit, ammeter, voltmeter, resistor, wire, rheostat*

1   electrons are moving A _____ the B _____

2   electrons are moving A _____ the B _____

3   electrons are moving A _____ the B _____

4   electrons are moving A _____ the B _____

5   electrons are moving A _____ the B _____

6   electrons are moving A _____ the B _____

7   electrons are moving A _____ the B _____

8   electrons are moving A _____ the B _____

ISBN: 9780170214667

Circuit design A

Date for completion: / /

Parent sig: _____
Teacher sig: _____

SP2 Unit 4.13–4.14

1 Design a circuit with a cell, a switch and a lamp, so that the light goes **off** when the switch is closed. Explain how closing the switch can turn the lamp off.

_____

_____

_____

_____

_____

_____

2 Draw a circuit diagram showing two lamps connected in parallel to a cell, with two switches in the circuit so each lamp can be turned on or off independently.

3 Draw a circuit diagram showing two lamps connected in series with three cells. Add one ammeter to measure the current in the whole circuit. Add one voltmeter to measure the energy of the cells. Add one fuse.

Physical Science

ISBN: 9780170214667

Physical Science

CHALLENGE

1 Draw a circuit diagram with a cell, three lamps (bulbs), and three switches so that each switch turns only one light on/off. Now add a fourth switch which could switch all three lights on/off if the other switches are closed (on).

2 Design a circuit with one cell, two switches and one lamp, so that the light comes on when either one of the switches is closed (on), or when both are closed. This circuit could be used to light the inside of a car that has two doors. The light is switched on when either of the doors is open.

24 **Current in series and parallel**

Date for completion: / /

Parent sig: _____
Teacher sig: _____

SP2 Unit 4.14

For each of the following circuits1 – 6, work out the current at each position where you see the symbols I, I_1, I_2 etc. (The symbol I for 'intensity' is often used to represent current.) In each case write your answer next to the = sign. Circuit 3 has been answered for you.

1

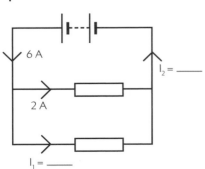

6 A

I_2 = _____

2 A

I_1 = _____

2

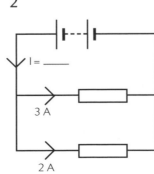

I = _____

3 A

2 A

3

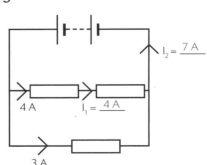

I_2 = __7 A__

4 A I_1 = __4 A__

3 A

4

> 6 A

> 2 A

> 1 A

$I_2 =$ ——

$I_1 =$ ——

5

↑ 4 A

> 1 A

> 2 A

$I_2 =$ ——

$I_1 =$ ——

6

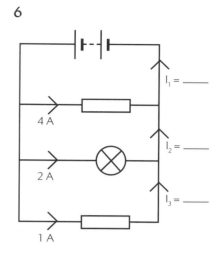

↑ $I_1 =$ ——

> 4 A

> 2 A

$I_2 =$ ——

$I_3 =$ ——

> 1 A

SP2 Unit 4.14–4.15

Physical Science

1 Complete these sentences: Current is the _____ of electrons

_____ a lamp or resistor in a circuit. Voltage is a measure of the

difference in _____ across a lamp or resistor or other component..

2 Write the current values (in amps) next to all the lamps (bulbs) in these circuits. The lamps are all the same type, and the battery is **12 volts** in each circuit.

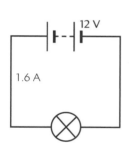

12 V

1.6 A

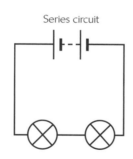

Series circuit

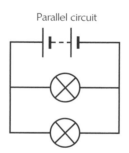

Parallel circuit

3 Write the current values (in amps) next to all the lamps (bulbs) in each of these circuits. The lamps are all the same type, and the battery is **6 volts** in each circuit.

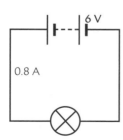

6 V

0.8 A

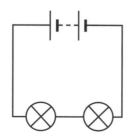

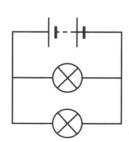

ISBN: 9780170214667

4 Write the current values (in amps) next to all the lamps (bulbs) in each of these circuits. The lamps are all the same type, and the battery is **3 volts** in each circuit.

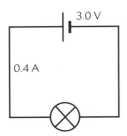 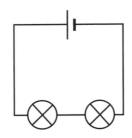 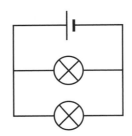

5 If one lamp in a series circuit is switched off, the

other lamp will _____.

If one lamp in the parallel circuit is switched off,

the other lamp will _____.

1 Complete the following sentences. The first letter(s) are provided.

Voltage can be the measure of energy su_____ by a cell or a

b_____; or it can be a measure of the amount of energy used up by a

l_____ or a r_____. Another name for the voltage is

po_____ d_____. The energy used or supplied by any

component in a circuit is measured by using a v_____. A voltmeter is always

connected in p_____ (across) to a component. A voltmeter must be

connected so that its terminals are connected positive to p_____ with the

battery or the power pack. Voltage is measured in v_____ (V). 1.0 volts is the

amount of e_____ that one coulomb of electrons will pick up from a source, or

else transfer to heat or light. The energy that a source (e.g. battery) provides will be all used up by

the other c_____ in a circuit.

Physical Science

2 Draw voltmeters in this circuit so as to measure the voltages across the cell and each lamp.

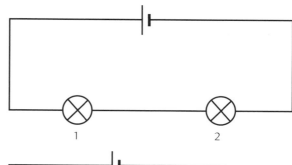

3 Kylie connects up this circuit. Her measurements of the voltages across the two lamps are shown.

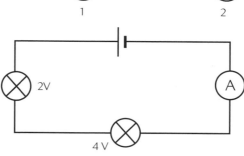

A What is the battery's voltage? _____

B Kylie notices that one light is brighter than the other. Which light? Explain why.

_____

_____

_____

_____

_____

C The current passing through the ammeter was 2 amps. Would Kylie have got the same reading if she had connected the ammeter **between** the lamps instead? Explain.

_____

_____

_____

_____

_____

_____

ISBN: 9780170214667

Physical Science

Complete the following sentences and the table using these 18 words and symbols. Keep uppercase and lowercase letters as given here: *kilogram, power, energy, W, V, s, J, millilitre, newton, physics, volt, amp(ere), coulomb, Systeme, m, K, N, A, L.*

More than any other branch of science, _____ depends on exact measurements. The basic units of measurement are known as **SI units**, short for '_____ Internationale'. This table has both SI and SI-related units of measurement.

What is being measured?	The basic unit of measurement	The unit symbol
time	second	
distance	metre	
mass		kg
volume	litre	
volume		mL
temperature	kelvin	
	joule	
force		
	watt	
electric current		
electric charge		C
electric energy, per coulomb		

1 Complete the sentences, choosing from among these words and symbols (some words may be used twice, some not at all): *kilowatt, volt, kW, energy, heat, watt, w, amp, joules, V, W, second.*

In science, 'power' means the amount of _____ that is converted

from one form to another every _____. The unit for power

is the _____, symbol _____, and also the

_____, symbol _____. Technically, power is measured

in how many _____ of energy converted per second. Power applies

to any kind of energy conversion. For example, a fluorescent light converts electric

_____ to light and also some _____.

2 People often use the word 'power' to describe anything electrical. The following statements do not use 'power' in the correct technical sense. Rewrite each sentence using terms like *energy, current* and *electric* in the correct places.

The power flows around the circuit from battery to bulb.

_____

A voltmeter measures the power of a lamp.

_____

Power is split up in a parallel circuit.

_____

That power heater is costing us a lot of money.

_____

3 The cost of using an electric appliance depends on three factors:
   • its power rating in kW
   • how long it is switched on for, in hours
   • the price of electric energy, measured in kWh (kilowatt hours). For the questions below, assume the cost is 25 cents per kWh.

$$\text{cost} = \text{power (kW)} \times \text{time (hours)} \times 25 \text{ cents}$$

A 3 kW water heater takes 2 hours to reheat after everyone in the family has had a shower. Calculate the total cost of hot water for showers each day for the family.

formula: _____

substitute: _____

calculate answer: _____

Physical Science

ISBN: 9780170214667

4 A 200 W computer is switched on for an average of five hours a day. Calculate the cost of electric energy used over one year (assume 25c per kWh).

formula: _____

substitute: _____

calculate answer: _____

5 A 2.5 kW heater is used for six hours a day for a week. Calculate the total energy cost.

formula: _____

substitute: _____

calculate answer: _____

29 Magnetic fields

Date for completion: / /

Parent sig: _____

Teacher sig: _____

SP2 Unit 4.17

1 Supply the six missing words choosing from: *opposite, electro, force, similar, horseshoe, permanent.*

There are two main types of magnets: _____ magnets, which includes

'bar' magnets made from a piece of flat metal, as well as _____

magnets that are curved into a U shape. Second main type: _____

-magnets that can be turned on or off. For all kinds of magnet, poles that are

_____ will attract each other, and _____ poles

repel each other. Any magnet is surrounded by a _____ field.

2 Draw the shape of the magnetic fields around these bar magnets, especially showing details of areas between magnets.

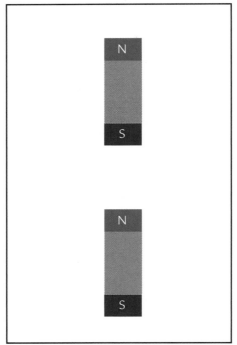

3 The inside of the Earth has a metal core that causes a huge magnetic field to surround Earth and nearby space. Draw this field. (The custom is that force field line arrows are drawn from N to S. Read *ScixncePlus 2* Unit 4.17 before you draw the arrows and label N and S on the Earth outline here.)

Date for completion: / /

Parent sig: _____
Teacher sig: _____

SP2 Unit 4.18

1 Only three elements can be used to make magnets: _____,

_____ and _____.

2 You have been asked by the local intermediate school to help their students make magnets. Write instructions on how to make a magnet in class. Make sure your instructions follow a logical order that any 11 year-old could follow. Include diagrams if you like.

_____

_____

_____

_____

_____

_____

_____

3 Magnetic levitation (or maglev) trains are extremely fast passenger vehicles, capable of speeds up to 581 km hr^{-1}. One of the newest is the Shanghai train which transports passengers 30 km to the airport in just 7 mins 20 seconds. German engineer Hermann Kemper was awarded the first patents for Maglev trains in 1937, but it wasn't until 1985 that the first passenger train was opened at Birmingham airport, England. It travelled at speeds up to 42 km hr^{-1}, but its track was only 600 m long. The 'levitation' part of the name suggests that the trains hover above the track,

and this is exactly right. The Birmingham train was raised 15 mm above the base of the track. This is achieved by magnets which repel each other inside the train, and in the walls of the track. This means that there is no need for steel tracks, the train is very quiet, and there is no energy lost as friction in the wheels. Electricity is used to control the strength of the magnets, and so different loads of people can be accommodated. The magnets are also used to propel the train forwards, as the magnets in front of the train can be used to attract the train to them, and the ones behind can repel it forward. Maglev trains are much more efficient than rail trains, although the cost of setting up a system has been quoted by several countries at $100 million US per km. However, maglev trains are much quieter, they work just as well in reverse as forward, and their schedules are less affected by weather.

Physical Science

ISBN: 9780170214667

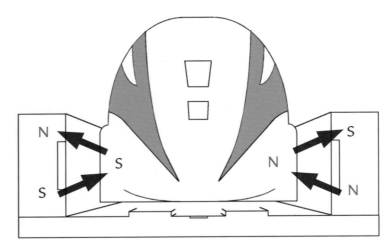

A What does 'maglev' stand for? _____

B The magnets that support the train also do another job. What is it?

_____

_____

C Calculate the Shanghai maglev's average speed.

_____

_____

D Suggest why it took more than 50 years before commercial maglev trains first operated.

_____

_____

E Suggest one advantage and two disadvantages of maglev trains.

_____

_____

_____

_____

31 Electromagnets

Date for completion: / /   Parent sig: _____
Teacher sig: _____

SP2 Unit 4.19

1 Complete the sentences below.

When electricity passes through a wire, a m_____ field is set up in the

area around the wire. Three factors affect the strength of this field: the strength of the

c_____, the number of c_____, and a metal

c_____. If the wire is made into coils, many loops combine to give a field

with a shape like the magnetic field around a _____ magnet.

2 Describe what is meant by a solenoid. Describe and draw a solenoid and its magnetic fields.

_____

_____

_____

_____

_____

_____

_____

3 These sentences explain how a loudspeaker works. Arrange the 'steps' A to F in a correct sequence that describes how electrical signals are converted into sound. Write the matching letters in the boxes below.

A The amplifier creates a current that is fed into the coil.

B The diaphragm stops the cone moving too far. This sometimes causes distortion if the current is too great.

C The amplifier then reverses, and pulls the cone back into the magnet

D The cone moves outwards, pushing air in front of it.

E The coil is repelled by the permanent magnet on the base.

F This is repeated hundreds of times a second to produce waves in the air that match the changing electrical current.

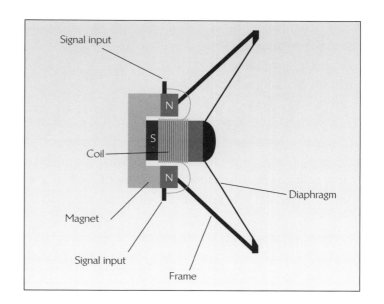

□ □ □ □ □ □

ISBN: 9780170214667

A relay is a switch in which a small electric current is used to turn on a device that uses a much bigger current. Examples of devices that use relay switches:

- starter motors in cars
- central locking switches in cars
- electric buzzers and bells.

A relay switch has two parts. The first part has a coil, shown here in blue. When you turn the coil on, it becomes magnetic, and attracts a metal component of the 2nd circuit, shown here in green. The green switch then closes (completes) the 2nd half of the circuit, moving the switch from position B to position C. This switches on an electrical component such as a starter motor. When the coil in the 1st part of the circuit is turned off, the coil stops being magnetic and a spring pulls the contacts apart.

1 Suggest what property (feature) of this type of magnet makes it suited to a relay switch.

_____

_____

2 Explain why steel (which is more than 80% iron) is suitable for the green switch.

_____

_____

3 Explain why it is necessary to have a spring attached to the green switch.

_____

_____

CHALLENGE

4 Suggest why cars use a relay switch to start, when you could just switch on the starter motor directly with a key. (Hints: think about what a key made of, and of the fact that a starter motor's current can exceed 150 amps.)

_____

_____

_____

_____

Physical Science

5 Add to the diagram, showing how and where a car starter motor and battery could be added. You do not have to draw details of the motor and battery, but make it clear how wiring should connect them to the relay switch.

33 Motors and generators

Date for completion: / /
Parent sig: _____
Teacher sig: _____

SP2 Unit 4.20

1 Complete this paragraph by choosing from these words: *kinetic, electrical, fixed, rotor.*

Electric motors and electric generators are similar in design. Both have a

_____(1), a moving part with many coils of wire; and are surrounded

by a _____(2) magnetic field. Motors and generators have

opposite purposes. A motor converts _____(3) energy to

_____(4) energy, and a generator converts

_____(5) energy to _____(6) energy.

2 In many situations it is possible to choose between electric motors and internal combustion (petrol) motors. Suggest four practical advantages of electric motors over petrol.

_____

_____

_____

_____

3 Suggest two advantages of petrol motors over electric – depending on the circumstances.

_____

_____

_____

_____

_____

_____

_____

_____

_____

ISBN: 9780170214667

4 A generator is a device that uses motion to create electricity. You can use a small electric motor for this purpose and turn it by hand, or you could make 100 watts of electricity if the motor/generator is connected to a bicycle. At peak times your household uses more than 10,000 watts on electricity, which would need 100 cyclists pedalling steadily. On a much bigger scale, giant generators – each as big as a house – can produce 250 million watts of electricity, also known as 250 megawatts (MW). The energy to spin the generators has to come from somewhere. In New Zealand, about 65% of all electricity is generated by the pressure of falling water, the biggest 'hydro' station being 750 MW at Manapouri. About 25% of generators are driven by steam that has been boiled by burning coal or gas. Example: Huntly can generate up to 1,000 MW. Geothermal heating is used at Ohaaki and Wairakei. 'Wind farms' that use wind's kinetic energy are growing in importance: some big turbines can generate up to 5 MW each. However, there will eventually be a limit to what can be generated in total, and we may have to settle for using less electricity.

Te Apiti wind farm.

Manapouri 'hydro' station.

Huntly power station.

Wairakei geothermal power station.

List any nine facts from the above, in bullet point form.

_____

_____

_____

_____

_____

_____

_____

_____

Generation technology

Electric energy can be generated in different ways. The main sources of energy for making electricity are coal, natural gas, hydro-electric (falling water), nuclear, tidal, wind, solar and bio-fuels. Almost all methods involve a generator – except for solar.

A **generator** is a big magnet (electromagnet rotor) made to spin past coils of wire. When the magnet moves past the coil it causes the electrons in the wires to move, and so electricity is made! A large energy supply is needed to make the big rotor move fast enough.

Here are three methods for generating electricity in New Zealand - coal, wind and water. Using these diagrams to help you, explain how each of the methods works. Do this in three separate paragraphs of 40-80 words each. In a final paragraph, summarise the main similarities and differences between the three methods.

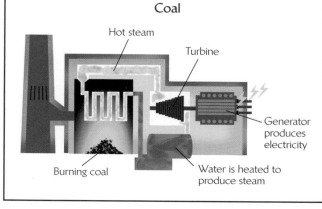

Coal

Hot steam
Turbine
Generator produces electricity
Burning coal
Water is heated to produce steam

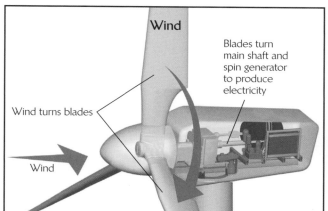

Wind

Blades turn main shaft and spin generator to produce electricity
Wind turns blades
Wind

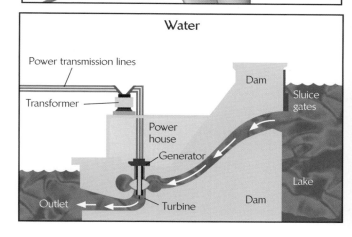

Water

Power transmission lines
Dam
Sluice gates
Transformer
Power house
Generator
Lake
Outlet
Turbine
Dam

Physical Science

_____
_____
_____
_____
_____
_____
_____
_____
_____
_____
_____
_____
_____
_____
_____
_____
_____
_____
_____
_____
_____

ISBN..9780170214667

Earth Science

1 Underneath the surface

1 Diagram 1 represents what probably exists beneath the Earth's surface. Choosing from the following seven labels, write each one in its correct place. *solid iron inner core; continental crust with lighter rocks; oceanic crust with denser rocks; convection currents in liquid magma; outer mantle; inner mantle.*

Diagram 1

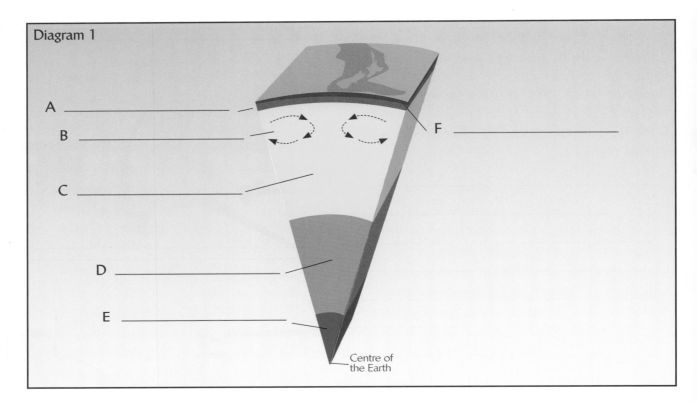

A _____

B _____

C _____

D _____

E _____

F _____

Centre of the Earth

2 Diagram 2 represents what is thought to exist beneath the North Island. The scale shows depth down to 100 km. For each lettered feature (A to E), write the word **evidence** or **inference**. See next page for word meanings.

A _____

B _____

C _____

D _____

E _____

Diagram 2

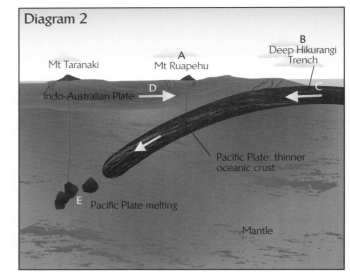

Mt Taranaki

A Mt Ruapehu

B Deep Hikurangi Trench

Indo-Australian Plate

D

C

Pacific Plate: thinner oceanic crust

E Pacific Plate melting

Mantle

Evidence, proof, prediction

SP2 Unit 5.1

Using the definitions in the box below, decide which category each of the following 17 statements falls into: **factual evidence**; **inference**; **prediction**; **proof**. Write your decisions in the second column.

Statement	Type
Deepest mines are hotter than shallow mines.	
At great depths, rocks are in a liquid state.	

Statement	Type
Japan has many earthquakes.	
Japan is on a plate boundary.	

Statement	Type
The Great Rift (GR) is a long wide valley.	
The GR has volcanoes and hot springs.	
The GR is a line where plates are moving apart.	
East Africa will split away from the rest of Africa.	

Statement	Type
Auckland has about 50 volcanoes.	
It is likely there will be more eruptions in the future.	
Auckland has an area of hot magma 50 km below.	

Statement	Type
Australia has no active volcanoes.	
Australia is not on a plate boundary.	

Statement	Type
Mid-Atlantic Ridge (MAR) rocks are volcanic.	
MAR rocks are younger than those either side.	
The MAR is volcanically active.	
Plates are moving away from the MAR.	

Evidence: facts or signs showing that something exists or is true.
Proof: evidence that is 100% convincing, or almost 100%.
Inference: something that you think is true, based on evidence.
Prediction: a statement about what you think is going to happen in the future.

Earth Science

ISBN: 9780170214667

NASA photograph

1 Write in the names of natural features A to J. Use an atlas if necessary.

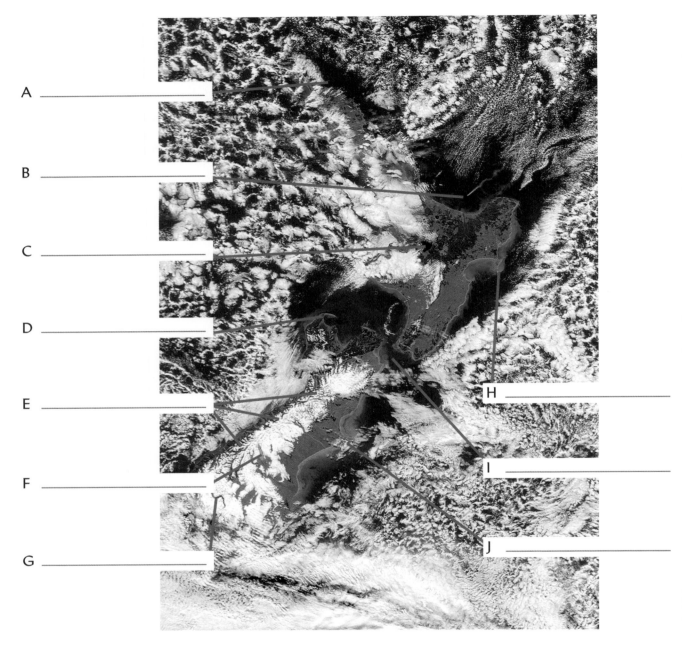

A _____

B _____

C _____

D _____

E _____

F _____

G _____

H _____

I _____

J _____

2 Give the letters that correspond to each of the following natural features:

- a major fault line associated with plate collision _____

- one active volcano _____

- three previously active volcanic areas _____; _____; _____; _____

- two lakes in valleys shaped by glaciers _____; _____

Zealandia

Date for completion: / /

Parent sig: _____
Teacher sig: _____

SP2 Unit 5.2

This map is coloured to show ocean of different depths in different colours. You may need an atlas to help you provide answers for the following tasks.

1 Use a pencil to mark the outline of the sub-continent 'Zealandia', which is presently more than 90% submerged.

2 Label as many islands as possible, including: New Caledonia, Samoa, Tonga, Fiji, Vanuatu, Niue, Chatham Islands, Campbell Island, Antipodes Island, Kermadec Islands.

3 Label other physical features such as Chatham ridge, Kermadec trench, Hikurangi trench.

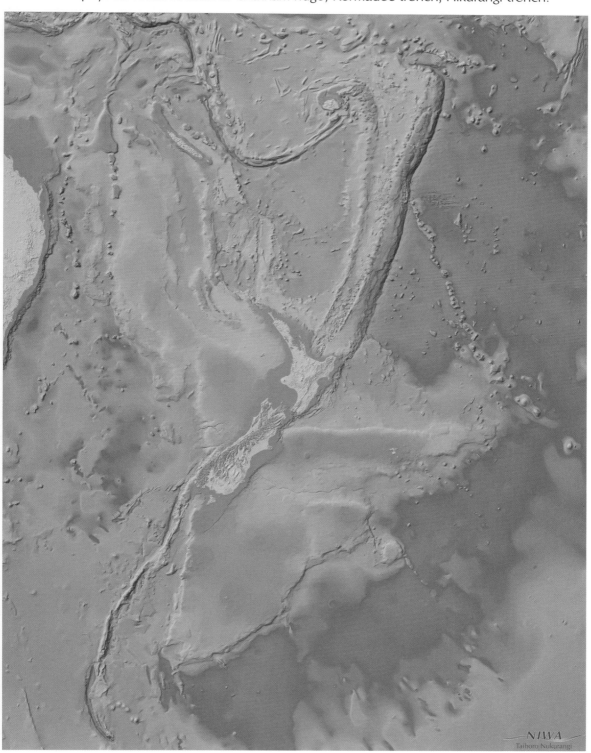

NIWA
Taihoro Nukurangi

Earth Science

ISBN: 9780170214667

Rock types

Date for completion: / /

Parent sig: _____
Teacher sig: _____

Add to this table by completing the middle column. Also write examples in the third column. Choose from this list of 12: *sandstone; limestone; granite; marble; pumice; pounamu (greenstone); obsidian; quartzite; shale; scoria; gneiss; basalt.*

Rock type	How this type originates	Some examples of this type
Plutonic igneous	From magma that has cooled slowly, deep below the surface.	
Volcanic igneous	From magma that has _____ _____ _____ _____ _____	
Sedimentary	From rocks that have _____ _____ _____ _____ _____	
Metamorphic	Rocks that have been changed by _____ _____ _____ _____	

Earth Science

Rock photos

Date for completion: / /   Parent sig: _____   Teacher sig: _____

State what kind of rock is shown in each of these photos. Also say what evidence each answer is based on. Choose from: *sedimentary (conglomerate)*, *sedimentary (sandstone)*, *metamorphic*, *granite*.

A

B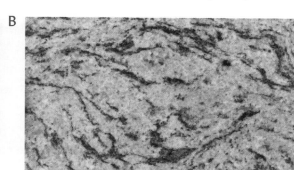

C

D

A Type _____

Evidence _____

_____

B Type _____

Evidence _____

_____

C Type _____

Evidence _____

_____

D Type _____

Evidence _____

_____

Earth Science

ISBN: 9780170214667

Date for completion: / /
Parent sig: _____
Teacher sig: _____

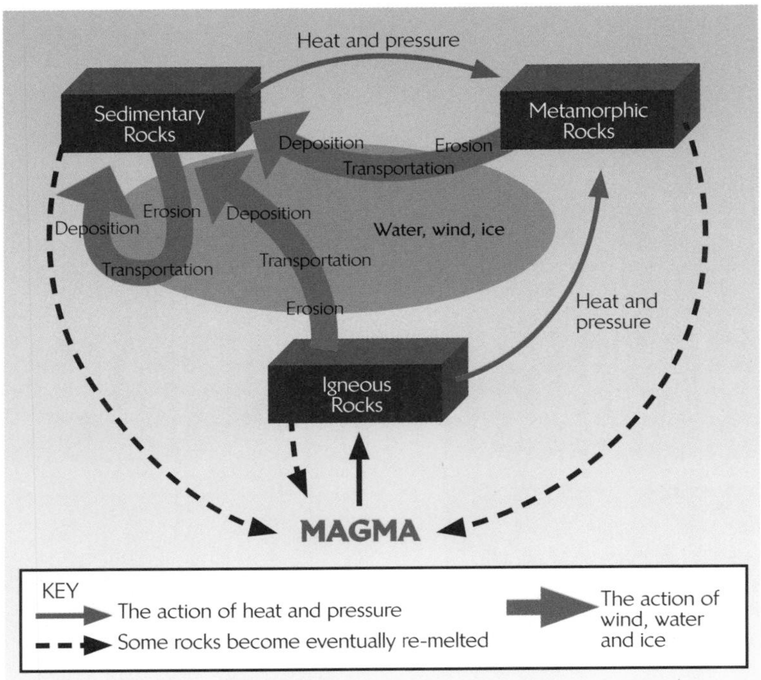

1 State which kinds of rocks are originally derived from magma: directly _____,

and which are derived from magma indirectly _____.

2 Name three processes that are always involved in the making of all sedimentary rocks:

_____, _____, and _____.

3 Name two processes that are always involved in the making of metamorphic rocks:

_____., and _____.

4 The diagram mentions transportation. Transportation of what and by what?

_____

_____

_____

5 Using information shown in the diagram, suggest why very old igneous rocks (more than 1 billion years old) are very scarce.

_____

_____

_____

_____

Earth Science

CHALLENGE

Measuring density

Date for completion: / /

Parent sig: _____
Teacher sig: _____

Density is the mass of one cubic centimetre of a substance, in g/cm^3. It is calculated by dividing the mass of an object by its volume, using the following formula:

$$\text{density} = \frac{\text{mass in grams}}{\text{volume in cm}^3} \qquad D = \frac{m}{v}$$

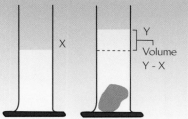

Maddison's aim was to find the density of several small rocks. She measured the volume of each rock by putting it under water and finding how much the water level rose. The results are in this table.

Rock	Mass (g)	Water volume before: X	Water volume after: Y	Y-X (cm³)	Calculate density: g/cm³
A	40.5	40	56		
B	66.0	70	93		
C	24.1	46	75		
D	372	29	44		
E	87.9	50	96		

(Note: 1 cm³ = 1mL. We generally use cm³ for solids, mL for liquids.)

1 Complete all columns of the results table. Round your density answers to two decimal places.

2 Predict which rock is likely to float in water _____
   Give a reason for your answer.

   _____

   Name what type of rock it probably is. _____

3 Identify which result is definitely wrong. _____
   Give your reason for deciding this result is wrong.

   _____

   _____

4 Suggest what kind of mistake Maddison made in getting the wrong result you identified above. Supply what was probably the correct result.

   _____

   _____

   _____

Earth Science

ISBN: 9780170214667

Complete the following sentences by writing in correct words selected from this list. Choose 14 from these 18: *craters, granite, pumice, geosphere, viscous, gas, rhyolitic, fertile, Krakatoa, steam, Taupo, caldera, ash, Rangitoto, lava, magma, basalt, solid.*

Some volcanoes have magma that results in free-flowing _____(1) that

hardens to form rocks such as _____(2). Other volcanoes have more

_____(3) magma. When this magma reaches the surface, the dissolved

_____(4) in it explodes violently, resulting in a _____(5)

eruption. Famous violent eruptions of this type include _____(6) and

_____(7). Rocks formed from this kind of explosive eruption include rhyolite,

and very low density _____(8), which is full of gas bubbles. A different kind of

explosive eruption can result from _____(9) caused by hot magma reaching

underground water. Several _____(10) around Auckland were formed by this

kind of explosion, and are probably linked to a hotspot _____(11) chamber

deep below the surface. A much bigger kind of crater, known as a _____(12),

results from the collapse of the entire volcano. Most eruptions also produced lots of gritty air-

fall material, generally known as _____(13). Soils eventually formed from

volcanic rocks tend to be very _____(14).

The map on the next page shows the epicentres and sizes of aftershocks in the Canterbury region, September 2010 to April 2011. Not every epicentre is visible, because circles for more recent events overlap some of the earlier ones.

Use the map to complete a tally of all the magnitude 3 to 6 earthquakes and aftershocks over this time period. As you record each circle, make a pencil mark on it to ensure you do not count it twice.

Suggestion: count in groups of five like this: ||||

Earth Science

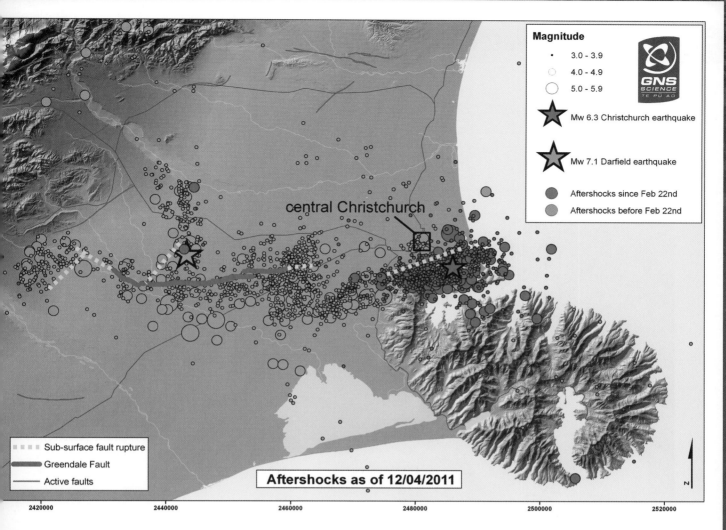

Magnitude

- • 3.0 - 3.9
- ○ 4.0 - 4.9
- ○ 5.0 - 5.9

★ Mw 6.3 Christchurch earthquake

★ Mw 7.1 Darfield earthquake

● Aftershocks since Feb 22nd

● Aftershocks before Feb 22nd

central Christchurch

▪▪▪▪ Sub-surface fault rupture

▬▬ Greendale Fault

—— Active faults

Aftershocks as of 12/04/2011

2420000   2440000   2460000   2480000   2500000   2520000

Magnitude	Tally	Total
3		
4		
5		
6	1	1
7	I	1

Earth Science

ISBN: 9780170214667

Date for completion: / /

Parent sig: _____

Teacher sig: _____

This is a valley below Aoraki/Mt Cook. This scene shows many features caused by ice and rivers cutting into the landscape. For each of the arrowed features A, B, C, D, explain how ice and/or running water caused that feature to form. Label features C and D.

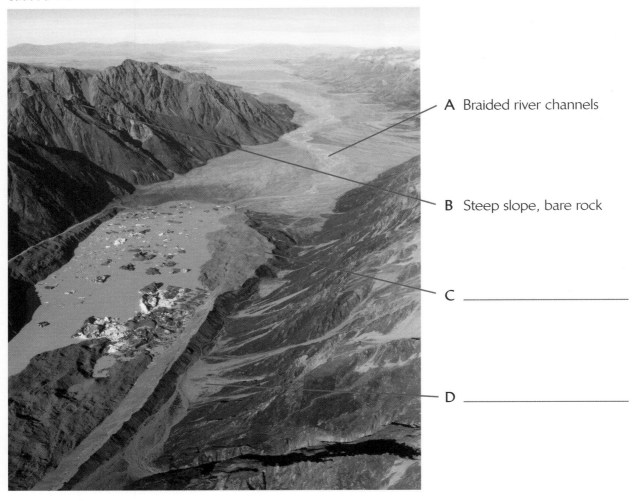

A Braided river channels

B Steep slope, bare rock

C _____

D _____

A _____

_____

_____

B _____

_____

_____

C _____

_____

_____

D _____

_____

_____

Earth Science

Date for completion: / /
Parent sig: _____
Teacher sig: _____

SP2 Unit 5.8/5.9

For 1 to 4, write the term (chosen from the list of four given in brackets after each statement) that will complete each sentence.

1 Water run-off from hills usually forms small _____
   (water cycles, drainage, streams, overflows)

2 A stream that flows down a steep valley is described as _____
   (mature, shallow, old, young)

3 A broad flat valley formed by a river is described as a _____
   (flood plain, drainage system, mature river, swamp)

4 Eroded rock deposited by a glacier is known as _____
   (landfill, new rock, glacium, moraine)

5 The diagrams show one river at different stages in the many thousands years of its history. Label each diagram with the word **young**, **mature**, or **old**.

 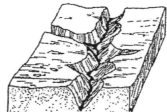 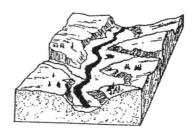

_____ _____ _____

6 Complete the following sentences by writing in words selected from this list: *chemical, erosion, pieces, acid, deposited, temperature, physical, limestone, soil, transported, roots, react.*

Weathering is the breaking of rocks into _____(1). There are two main

types: _____(2) weathering causes rocks to be cracked by extremes

of _____(3), and even by the _____(4) of

plants. Secondly, _____(5) weathering involves water, air and other

substances which _____(6) with minerals in the rocks. In the case of

_____(7), CO_2 in the air forms an _____(8)

solution that dissolves the rock. Whatever kind of weathering takes place, the result is

loose _____(9) or clay or gravel of one kind or another. The removal

of loose material is known as _____(10), and this material is sooner

or later _____(11) over long distances by wind or running water and

_____(12) elsewhere.

Earth Science

ISBN: 9780170214667

SP2 Unit 5.10

A	Quartz veins develop inside the Greenland Group rocks.
B	Uplift and erosion causes the landscape to take on its present shape.
C	A long period of erosion smooths the land surface above HC and GG.
D	Greenland Group rocks become folded.
E	More sediments are laid down directly above the first coal strata.
F	A geologist surveys the rocks and draws the cross-section.
G	New sedimentary phase results in coal strata being laid down.
H	Greenland Group sedimentary rocks laid down.
I	Movements on a fault cause HC to be positioned alongside GG.
J	Second and third coal strata are deposited.

This diagram shows a cross-section of the Brunner coalfield. GG is short for the Greenland Group rocks, HC is short for Hawk's Crag rocks. The 10 events A to J are listed in the wrong order.

1   Explain why event A must have happened after event D.

_____

_____

2   Write all ten events in the correct order, from oldest to most recent.

_____

_____

_____

_____

_____

_____

_____

_____

_____

_____

3   Draw a cross-section of the landscape to show what was probably happening at stage C.

Earth Science

CHALLENGE

Recycled water

Date for completion: / / Parent sig: _____ Teacher sig: _____

The diagram below shows a simple version of the water cycle. Write the following eight words and phrases in their correct places as labels on the diagram. Some words may be used more than once: *evaporation; groundwater flow; precipitation: rain and snow; ocean; condensation; surface runoff; transpiration from plants.*

Earth Science

ISBN: 9780170214667

15 Tech word review

SP2 Unit 5.1–5.12

Date for completion: / /

Parent sig: _____
Teacher sig: _____

In the second column, write the technical words that match each of the descriptions in the left column. Choose from this list of 20: *ice; Taupo; aquifer; isotopes; subduction; Himalayas; caldera; anticline; moraine; Gondwana; Richter; volcanic; stack; Zealandia; minerals; fiord; rhyolitic; braided; syncline; seismometer.*

Definition or description	Technical word
The southern super-continent that existed 200 million years ago	
Name of the subcontinent that includes present-day New Zealand	
Region where one tectonic plate is being pushed under another	
Mountain range caused by India colliding with Asia	
Any wide sunken area caused by a collapsed volcano	
Most rocks consist of several of these joined together	
Volcanic explosions from magma with lots of dissolved gas	
An upward fold in rock strata	
A downward fold in rock strata	
Pile of shattered rock pushed aside by a glacier	
Sea filling a glacier gouged out by glaciers	
A pillar of rock left behind when sea erodes coastal cliffs	
A river that has many smaller interconnected channels	
A big volume of underground water	
77% of the world's fresh water is in this form	
Forms of an element with different numbers of neutrons in the nucleus	
Earthquake intensity scale	
Instrument that measures earthquake strength	
Site of New Zealand's biggest eruption in the past 20,000 years	
Igneous rocks from magma that cools and hardens on the surface	

SP2 Unit 5.12

Planet Earth was formed around 4,500 million years ago (mya). This age is based on good evidence, but it's almost impossible to imagine even one million years. One way to visualise vast time-spans is to imagine a scale where 1 mm = 1 year. Using this scale, convert each of the events below into a distance measurement.

For any distance over 1,000 mm, write this in metres. For anything over 1,000 metres, write your answer in kilometres. For events more than about 3,000 years ago dates are approximate (**mya** is short for 'millions of years ago'; **ya** for 'years ago').

Event	Years in the past (or actual date)	Distance in the past, on scale 1mm = 1yr
your birth	( )	
outbreak of World War 2	(1939)	
treaty of Waitangi signed	(1840)	
birth of Galileo & Shakespeare	(1564)	
earliest evidence of writing	about 7,000 ya	7 metres
most recent ice age ended	about 14,000 ya	
humans first arrive in Australia	about 40,000 ya	
humans first begin to leave Africa	about 70,000 ya	
first *Homo sapiens* humans	about 400,000 ya	
pre-humans first begin to use fire	about 1 mya	
tectonic plate collision begins under NZ	about 20 mya	20 km
mammals start to become common	about 55 mya	
extinction of the dinosaurs	about 60 mya	
breakup of Gondwana is underway	about 180 mya	
rise of the dinosaurs	about 250 mya	
first vertebrate animals on land	about 360 mya	
first plants on land	about 420 mya	
first vertebrate animals exist in the sea	about 500 mya	
earliest evidence of life	about 3,500 mya	
planet Earth is formed	about 4,700 mya	
origins of the universe ?	about 13,000 mya	

Earth Science

ISBN: 9780170214667

Astronomy

1 Galaxy distances

SP2 Unit 5.13

Date for completion: / /

Parent sig: _____
Teacher sig: _____

Galaxy	Distance from us in light years	Angle in degrees	Type of galaxy	scale distance: 4 cm = 1 million light years
IC10	2,690,000	119	Irregular	2.69 x 4 = **10.7 cm**
M31	2,560,000	121	Spiral	
M33	2,735,000	133	Spiral	
Sagittarius Dwarf	78,000	6	Elliptical	
Leo A	2,250,000	197	Irregular	
Sculptor Dwarf	258,000	288	Elliptical	
Carina Dwarf	329,000	260	Elliptical	
Phoenix Dwarf	1,450,000	272	Irregular	1.45 x 4 = **5.8 cm**
NGC 6822	1,520,000	25	Irregular	
Large Magellanic Cloud	165,000	281	Irregular	

The table gives distances and directions of galaxies near our own galaxy, the Milky Way.

1 Convert real distances to scale distances in the column on the right.

2 Use the table information to plot the exact positions of all 10 galaxies in a diagram on the next page. Use a protractor to get the angles. Label each position with the galaxy's name. The Milky Way is at the intersection of the two lines.

3 This method and your scaled star map has at least one important defect. Identify the defect.

_____

_____

_____

_____

CHALLENGE

Solar System

0°

270°

90°

180°

ISBN: 9780170214667

Astronomy

SP2 Unit 5.14

Date for completion: / /

Parent sig: _____

Teacher sig: _____

The constellation 'Crux' helps boat skippers and mountaineers find their way at night. Here's how.

On this photograph of the night sky:

1 Draw two circles, one around the Southern Cross and one around the pointers. Write the name of these two groups at the side.

2 Draw a long line 'A' through the central axis of the Southern Cross.

3 Draw a long line 'B' that is a perpendicular bisector of the short line between the two pointers.

4 Where A and B meet, mark 'South Celestial Pole (SCP)'

5 Mark the direction 'South' on the horizon.

ISBN: 9780170214667

Date for completion: / /   Parent sig: _____   Teacher sig: _____

SP2 Unit 5.13–5.20

The *Cassini* unmanned space probe was launched in 1997. Its mission: to study and explore Saturn and its many moons. It arrived near Saturn in 2004, and over the next four years sent back masses of dramatic new information to teams of scientists in many countries.

More moons of greater variety orbit Saturn than any other planet. 18 were known at the start of the mission, and *Cassini* found dozens more. These natural satellites range from Titan, nearly the size of Mars, down to tiny moonlets less than 1 km across.

Some of the discoveries include ice geysers shooting from Saturn's moon Enceladus, and confirming that one of Saturn's rings is created from these ice particles. An onboard radar instrument, which sees through clouds, unveiled the fascinating world of Titan. Titan ripples with mountains, the highest peaks rise up to about two kilometres.

The Huygens probe, built and managed by the European Space Agency, was bolted to *Cassini* for its journey to Saturn. Huygens' descent to Titan in 2005 marked humankind's first attempt to land a probe on another world in the outer solar system. The probe continued to send data for about 90 minutes after reaching the surface.

Huygens captured the most attention for providing the first view from inside Titan's atmosphere and on its surface. The pictures of river-like channels and rounded ice boulders surprised scientists with the extent of the moon's similarity to Earth. They showed evidence of lakes of methane, with ethane rain, and erosion from running liquids.

Because of the very dim sunlight at Saturn's orbit, solar arrays are not feasible. Electrical power is supplied to the orbiter by a set of radioisotope heat generators, which convert the heat from the natural decay of plutonium-238.

The Cassini-Huygens mission was a cooperative project between NASA, the European Space Agency, and the Italian Space Agency. The hugely complex probe weighs two tonnes, and the whole project cost over $3 billion. (Information from www.nasa.gov/cassini)

From each of the seven paragraphs above, write 2 or 3 key items of information in the form of short bullet points.

_____

_____

_____

_____

_____

_____

Astronomy

ISBN: 9780170214667

_____

_____

_____

_____

_____

_____

_____

_____

_____

_____

_____

_____

_____

_____

_____

_____

_____

_____

_____

4   Star life and death

Date for completion: / /

Parent sig: _____
Teacher sig: _____

SP2 Unit 5.16/5.17

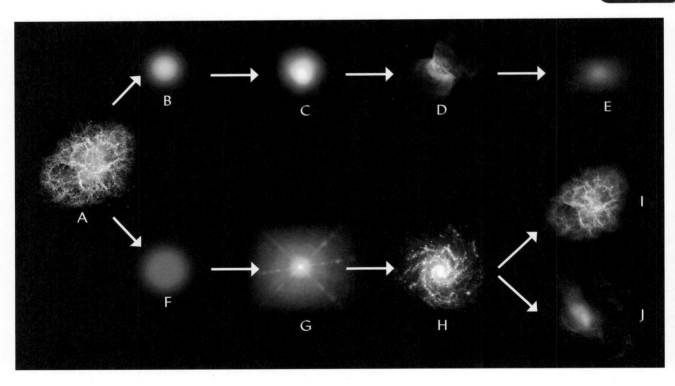

The diagram above simplifies the life cycle of stars. Massive stars have a different development to average and smaller stars (such as our Sun).

1 Using the following terms, provide descriptive names for stages A to J in the diagram above: *Red supergiant; Planetary nebula; Average star; Massive star; Black hole; Stellar nebula; White dwarf; Neutron star; Red giant; Supernova.*

A _____   F _____

B _____   G _____

C _____   H _____

D _____   I _____

E _____   J _____

Stars vary in colour, temperature, and mass. This table gives the physical properties of 12 different stars.

Name	Temperature (kelvin)	Colour	Star mass
Sirius	25,200	Blue	2.02
Beta Procyon	6,530	White	1.42
Sol (our Sun)	5,788	Yellow	1.00
Betelgeuse	3,500	Red	19.00
Bellatrix	21,500	Blue	9.00
Alpha Centauri A	5,260	Yellow	1.10
14 Herculis	5,000	Orange	0.79
Polaris	7,200	White	7.55
Vega	9,620	White	2.10
Rigel	11,000	Blue	17.00
Epsilon Indi	4,630	Orange	0.76
Antares	3,388	Red	18.00

2 Rearrange these stars in order from hottest to least hot.

Name	Temperature (kelvin)	Colour	Star mass

3 From this information, calculate the temperature range for each colour.

Blue _____   White _____   Yellow _____   Orange _____   Red _____

ISBN: 9780170214667

Astronomy

4 Calculate the average solar mass for each colour.

Blue   White   Yellow   Orange   Red

_____   _____   _____   _____   _____

5 Our Sun is using up its hydrogen. According to evidence, about how long will the Sun last before becoming a red giant?

_____

5   Space summary

Date for completion: / /   Parent sig: _____   Teacher sig: _____

SP2 Unit 5.13–5.20

Supply the missing words, choosing from this word list (each word is used once): *reflects; nuclear; gravitational; Mercury; nebulae; Jupiter; year; galaxy; orbit; rock; gas; galaxy.*

There are eight planets in the solar system. All of these planets _____(1) around the Sun. The largest is _____ (2) and the smallest is

_____ (3). The time it takes for a planet to orbit the Sun is

known as one _____ (4). The outer surfaces of Mercury, Venus,

and Mars are mostly _____ (5), while Jupiter and Saturn are

mainly _____ (6). The Sun generates heat and light energy

because of _____ (7) reactions inside it. Some of this light

_____ (8) off the planets, which is why we can see them on a

clear night. The Milky Way is a _____ (9) which is made up of stars,

_____ (10) and other celestial bodies, which stay together because of

_____ (11) forces.

6   Astronomy crossword

Date for completion: / /   Parent sig: _____   Teacher sig: _____

SP2 Unit 5.13–5.20

1 Choose any 9 to 12 words from the word list below, and use them to create your own crossword in the grid provided on the next page. Outline the squares for your words and give each word a number. Then, write your own numbered clues for your chosen words.

nebula	galaxy	constellation	Orion
Crux	zodiac	Hubble	bigbang
cosmology	helium	hydrogen	supernova
neutron	milkyway	Galileo	Titan
pole star	parallax	red shift	Matariki

2 Exchange a blank crossword grid and your written clues with someone else in the class. Test each other.

CLUES:

Across:

Down:

ISBN: 9780170214667

The Kepler Mission is a NASA Discovery Program for detecting potentially life-supporting planets around other stars. All the extra-solar planets detected so far by other projects are giant planets, most are the bigger than Jupiter. *Kepler* is poised to find planets as little as 1/600 the mass of Jupiter, and has already discovered a growing number of planets.

How does it work? When we see a planet pass in front of its parent star, it blocks a small fraction of the light from that star. When that happens, we say that the planet is transiting the star. If *Kepler* sees repeated transits at regular times, it has discovered a planet! From the brightness change we can tell the planet size. From the time between transits, we can tell the size of the planet's orbit and estimate the planet's temperature. These qualities determine possibilities for life on the planet.

The *Kepler* spacecraft has a 0.95m diameter telescope with 40 two-megapixel CCD light sensors. It is designed to continuously monitor 100,000 of the brighter stars in the constellations Cygnus and Lyrae – an area about the size of your open hand held at arm's length.

To detect an Earth-size planet, the photometer must be able to sense a drop in brightness of only 0.01 %. This is like sensing the drop in brightness of a car's headlight when a fruit fly moves in front of it! The photometer must be space-based to obtain this precision. *Kepler* was launched in March 2009, and orbits the Sun in Earth's orbit, but trailing millions of kilometres behind Earth.

Data is stored onboard, and sent back about once a week.

The spacecraft is named after Johannes Kepler, born in 1571 to a very poor family. As a professional astronomer, Johannes developed precise calculations on the movements of plants that are used to this day. He predicted exactly the transit of Mercury on 9 November 1769 that was a major reason for Captain Cook's voyage of discovery to New Zealand. (Information from kepler.nasa.gov)

Kepler spacecraft being prepared for launch.

From each of the five paragraphs above, write three or four key items of information in short bullet form. Your total a number of bullet points should be no more than 20 lines.

_____

_____

_____

Solar System

ISBN: 9780170214667

Glossary

acceleration   the rate of change of speed or of velocity

accretion disc   a disc of gas and dust rotating around a star or black hole

acid   a compound that releases hydrogen ions when it dissolves in water, and has a pH of less than 7

adaptable   able to survive and be successful in changing conditions

adaptation   in evolution, any feature or structure of an organism that makes it well-suited to its environment

addictive   a substance that causes the user to become chemically or psychologically addicted to it

alcohol   an organic compound with at least one – OH group joined to a carbon atom

alkali   a soluble base, with pH greater than 7

alleles   different forms of one gene

allergic   an abnormal reaction of the body to a substance, often causing itchy eyes, runny nose, wheezing, skin rash

allotrope   structurally different form of the same element eg. diamond and graphite are allotropes of carbon

alloy   a mixture of different metal elements

amp   the unit in which electrical current is measured. Symbol A

anaesthetic   a chemical that temporarily reduces sensory messages to the brain, usually to stop pain

anhydrous   without water. Anhydrous chemicals can be used as drying agents

antibiotic   a fungus-derived substance that kills bacteria, or limits their growth

antibodies   immune-system proteins that attack 'foreign' material

anticline   an upward fold in rock strata

antigen   any substance or material, such as the surface of a virus, that triggers an antibody response

aquifer   any big volume of underground water

ash   gritty particles of volcanic rock less than a few mm in diameter, given out during an eruption

atmosphere   the 'blanket' of gases around a planet

atomic mass   a measure of how massive an atom is, on a scale in which hydrogen atoms are mass 1

atomic number   the number of protons in the nucleus of the atom of an element.

bacteria   very small one-celled organisms that live in air, earth, water, plants and animals. Most kinds are saprophytes ie. agents of decay. Some kinds cause disease

baking powder   a mixture of tartaric acid and sodium bicarbonate

baking soda   sodium bicarbonate

basalt   igneous rock formed from lava having a lower silica content. Some kinds of basalt are very hard, others are full of gas bubbles (eg. scoria)

base   (1) in chemistry, a substance with a high pH, usually containing an -OH group. (2) in DNA, a molecule that is part of the genetic code. There are four DNA bases, known by the letters ATCG

Benedict's test   a chemical test used to show the presence of sugars

Big Bang   a mathematical theory for the origin, evolution and structure of the universe

binary fission   a form of non-sexual reproduction in which a living cell divides in two

biomass   how many tonnes of living things there are per square kilometre

biotechnology   the use of living things, especially bacteria, in industrial processes

black dwarf   a small, cold dead star

black hole   a star so dense that not even light can escape from its surface

caldera   a wide crater formed by the collapse of a volcano

carbohydrate   a compound made up of only C, H and O, with H and O always in the ratio 2:1. An important food group that gives energy eg. starch and sugars

carnivore   any animal that eats other animals

cell wall   thick outermost layer of all plant cells

cell   (1) tiny unit of life. (2) device for providing electrical energy

cellular respiration   the release of energy from food. May be shortened to **cell respiration**

cellulose   a tough carbohydrate, occurring in the cell walls of plants, indigestible by humans

characteristics   distinctive features, including how an organism is built, how it behaves, and how it functions

chemical bonds   the forces of attraction between atoms or ions

chemical formula   a shorthand way of showing which elements are present in a substance, and the relative numbers of atoms

chemical reactivity   the ability of a substance to react with other substances

chlorophyll   green substance in plants that traps light energy for photosynthesis

chromosome   a structure in the nucleus that contains many genes. Made mostly of DNA

clone   a group of organisms that all have identical DNA

cluster   in astronomy, a group of galaxies that are bound together by gravitational forces

codon   three DNA bases in sequence. Each codon 'signals' one particular amino acid

community   all the living things in a particular place. A community includes plants, animals, fungi and bacteria

component   a device in an electric circuit designed to convert electrical energy into other forms of energy

compound   a chemical substance made from two or more elements combined together

concentration   (of a solution) a measure of how much of a substance is dissolved in each litre of a solution

conclusion   what the results suggest or tell you

conductor   any substance which allows heat or electricity to flow though it easily

constellation   a group of stars that form a pattern when seen from Earth

consumer   any organism that feeds on other organisms. Includes all animals, fungi, bacteria

control   the version of an experiment that does not include the factor you are testing, in order to provide a basis for comparison

conventional current   the flow of electrical current from a positive terminal to a negative terminal

cosmology   the study of the origin, evolution, and structure of the universe

covalent bond   a chemical bond that involves sharing electrons

crust   the outer solid rocky layer of the Earth, from 10 km to 100 km thick

current   the flow of electric charge through a conducting material, usually electrons through metal wires

dark matter   a form of matter that can not be seen but is known to exist because of the way it affects visible matter

data   any known information, whether from observation or experiment: particularly measurements, but also information in word or picture form

decomposer   organisms like bacteria or fungi, that break down and decay the cells of dead plants and animals

density   the mass of one cubic centimetre of a substance (g/cm^3). It is calculated by dividing the mass of an object by its volume

dependent variable   the variable in an experiment or trial that provides the results. 'Dependent' because it may depend on factor X

digestion   the chemical breakdown of big molecules into smaller ones

diploid   a full set of chromosomes. Always an even number

discharge   the loss or removal of built-up static electrical charge

dissolves   mixes into water, or another solvent, forming a solution

DNA   a giant thread-like double-helix molecule that carries genetic information in coded form

dominant   in genetics, an allele which shows its effect when there is only one allele of a pair

ecology   the study of the interactions between living things and their environment, usually in a natural setting

ecosystem   a living community together with its physical surroundings. The main components of an ecosystem are plants, animals, decomposers, sunlight, recycling chemicals, and water. Intact ecosystems tend to be stable, and self-sustaining, and self-regulating

eggs   female gametes; ova

electric circuit   a complete path around which electricity flows

electrolyte   a substance that contains free ions, and so conducts electricity

electromagnet   a device consisting of a coil of wire wrapped around an iron core, and carrying an electrical current. The strength of the magnetic field can be adjusted or turned off

electromagnetic radiation   a form of energy that includes ultraviolet, infrared, X-rays, light and radio waves. All travel at 300,000 km per second

electrons   tiny negatively-charged particles that spin around the nucleus

ellipse   a particular geometric shape (eg. of the Earth's orbit around the Sun); not an oval

empirical method   scientific method based on observations, measurements, tests, and experimental results

energy transformation   the changing of energy from one form to another, such as from electric energy to light energy

enzyme   chemical substances produced by living cells, and which influence the speed of chemical reactions, while not being changed themselves

evidence   facts or signs showing that something exists or is true

evolution   the process of genetic changes in plants and animals over long periods of time. Also known as 'descent with modification'

exponential   a rate of increase that becomes quicker and quicker, repeatedly doubling numbers within the same time interval

fair test   a test in which all variables except for one are kept the same

fault(line)   long line in the Earth's rocks, where a mass of rock has fractured and the two sides are moving in opposite directions. Movement may be vertical, sideways, or a combination

fermentation   process by which sugars are converted to alcohol and carbon dioxide by organisms such as yeast organisms to obtain their energy

fertilisation   (or **conception**) joining of male and female sex cell. (gametes), producing a new single cell (zygote) that contains the DNA of both parents

filtrate   the liquid that passes through filter paper during filtration

flagella   microscopic whip-like structures that some microorganisms use to swim with

foetus (fetus)   the name given to the developing child after about 6–8 weeks. Before this: embryo

food pyramid   a diagram that shows the total biomass at each level in a food chain

force   a push, or a pull, or a twist

fossil   the dead remains of an animal or plant in which the bone (or other substance) has been replaced by mineral crystals

fossil fuel   any burnable substance formed over millions of years from dead plants and animals

friction   a force which opposes movement

fulcrum   a turning point for a lever. Also known as a pivot or a hinge

fungus   plant-like organisms that do not contain chlorophyll. Most kinds are saprophytes

galaxy   a large collection of stars, nebulae and other celestial bodies bound together by gravitational forces

gametes   sex cells (eggs and sperm). These only contain half the DNA present in body cells

gamma rays   very high energy short-wavelength electromagnetic radiation. Dangerous to living things

gas exchange   the diffusion of oxygen in one direction and of carbon dioxide in the other direction

gene   a length of DNA, in most cases with one gene carrying the code for one particular protein

genetic engineering   technologies by which new genetic combinations are artificially created, often by using DNA from different species. Also known as genetic modification (GM)

genome   the total genetic information carried by an organism

genotype   gene makeup, written in ways such as BB or Bb

glacier   a slow-moving river of ice

global warming   the enhanced greenhouse effect, in which increased amounts of gases like carbon dioxide trap extra heat by preventing it from being radiated into space, and in this way raise the temperature of the planet

glucose   simple 6-carbon monosaccharide sugar molecule common in plants.

Gondwana   the huge southern continent that began to break up 200 million years ago

gravity   a force of attraction between all things that have mass

group   a vertical column of elements in the Periodic Table. Elements in the same group have similar chemical properties

habitat   the usual place or 'address' of an ecological community, or of a particular species

haemoglobin (hemoglobin)   red protein in red blood cells. Carries oxygen

haploid   half of a full set of chromosomes; as in gametes

heat   form of energy related to the speed of particle movement

herbivore   any animal that eats plants

hormone   chemical released by cells, and affecting cells in other parts of the body

hydrocarbon   a compound containing hydrogen and carbon atoms only

hydrophilic   water-loving

hydrophobic   water-hating

hydrosphere   the layer of water that covers most of the Earth, in sea, lakes, aquifers, and ice

hyphae (hi-fee)   the individual feeding-threads of a fungus

hypothesis   a suggestion or possible explanation that often includes the words 'If… then…'

igneous   rock formed from solidified magma

immunise   to stimulate the body's natural defense immune response in order to build resistance to a specific infection

independent variable   The 'factor X' variable that you are investigating in an experiment or trial

indicator   a substance that can be used to determine the pH of a substance

induction   the process of a surface or object becoming electrically charged by bringing another charged object close to it

inert   chemically unreactive under normal conditions

ISBN: 9780170214667

inertia the tendency of any mass to resist movement or a change in speed

inference something that you think is true, based on evidence

infertile unable to reproduce; sterile

informed opinion an opinion based on knowledge and evidence

inorganic a substance that does not contain carbon atoms. (Except for CO_2, CO and carbonates which are inorganic)

insulator an object which doesn't allow electric charge or heat to flow through it

insulin a hormone involved in the regulation of blood sugar

ionic bond a chemical bond resulting from the attraction between positively and negatively charged ions

ionic compound a compound with ions held together by ionic bonds, usually between a positive metal ion and a negative non-metal ion

ion atom with an electrical charge that has resulted from the loss or gain of one or more outer electrons

isotopes forms of an element with different numbers of neutrons in the nucleus. Isotopes have the same atomic number, but different atomic masses

kaitiaki Maori principle of guardianship, and of caring for resources

kelvin scale the SI temperature scale of temperature. Unit: K

kilojoule (kJ) unit of energy

kilowatt-hours the unit in which total quantities of electrical power is bought and sold. Symbol kWh

LED light emitting diode

lever a simple machine, in the form of a long bar that reduces the amount of force needed to move an object

light year the distance light travels in a vacuum in one year

limestone sedimentary rock with a high calcium carbonate content

limewater calcium hydroxide solution; used to test for the presence of CO_2. Turns milky when carbon dioxide bubbles through it. Formula $Ca(OH)_2$

lipids fats and oils from animals and plants

lithosphere the layer of the Earth that includes the crust and the outer mantle

litmus one kind of indicator used to show whether a substance is acidic or basic

low concentration few particles in a given volume

lustre shininess, a property of metals

magnetic field a region where a magnetic particle experiences a force

malleable can be hammered into shape without breaking

mantle the semi-fluid layer of the Earth underneath the crust

mass a measure of the amount of substance in an object. Mass is measured in kilograms

Matariki Maori name for a group of stars also known as the Pleiades

mauri Maori concept of the energy that binds all things together, including the understanding that humans are part of the natural world

meiosis a special kind of cell division that produces eggs and sperm, and halves the number of chromosomes

metals chemical elements that are shiny, good conductors of electricity, and tend to give away electrons

Milky Way the galaxy that includes the Sun

mitosis the process of ordinary cell division

model in science, usually an overall picture or explanation or analogy of how a system works. Example: ecosystem model

molecule a group of atoms joined together

monomers molecular building blocks which are joined together to make a polymer

moraine a pile of jumbled and shattered rock that has been moved by a glacier. If at the end: a terminal moraine. If at the side: a lateral moraine

mould a growth of fungi, often as a downy or furry coating

MRSA a type of bacterium that Is resistant to most antibiotics. Short for methicillin-resistant *Staphylococcus aureus*

mtDNA a form of DNA which is inherited only from mothers, and is passed on only down the female side of the family

mutation an accidental alteration in a genetic message. A mutation may be harmful or helpful, small or big

mutualism an association between two different species, to the benefit of both

mycelium (*my-seely-um*) a mass of hyphae in a fungus

mycorrhiza mutualistic association of a fungus and plant roots

natural selection a process which results in some individuals having more offspring surviving to adult reproductive age

nebula A large, tenuous cloud of gas and dust in space. The gas is mostly diatomic hydrogen

negative feedback any sequence of responses that results in the opposite effect to whatever it was that triggered the initial response. A negative feedback process tends to produce an almost-stable equilibrium

neutron star a very small, extremely dense star that consists mostly of neutrons very, very close to one another

Newton's laws Newton formulated a number of laws, but 'Newton's laws' usually refers to his three laws of motion which describe how objects move when forces are applied

niche an animal's 'occupation', which includes its adaptations to feed and survive in a particular community

observation something you see, feel, hear, smell or taste

organic a carbon-containing substance produced by living things

organism any living thing

osmosis a special case of diffusion which occurs when two different solutions are separated by a permeable membrane. Water moves from the side in which there is more water, to the side in which there is less water

ovary the organ in which eggs are produced

ovum (egg) female gamete

oxidation any reaction between a substance and oxygen

ozone a form of oxygen; O_3

parallel a method for connecting components 'side by side' in an electrical circuit. There is more than one path for current to flow through

parasite animal, plant, fungus, bacterium or virus that lives and feeds on or inside another living thing

pathogen any disease-causing organism

periodic table a table in which the elements are listed in order of increasing atomic number, and elements with similar properties are in the same vertical column (group)

pH scale a scale from 1-14 used to describe how acidic or basic a solution is

phenotype the 'physical' result of genes, such as blue eyes or red hair

photosynthesis chemical reaction in green plants, in which the plant uses light energy to convert carbon dioxide and water into glucose and oxygen

plankton any floating, drifting, or weak-swimming organisms that live in water

plutonic igneous rock formed from magma that has cooled and solidified deep below the surface eg. granite

polar molecule a molecule with one part carrying a surplus +ve charge, and one part carrying a –ve charge

pollination process of transferring pollen from one flower to another flower of the same species

polymer a large chain-like molecule made up of repeating units of monomers

population the total number of a particular animal species in a particular area

positive feedback a sequence of responses that leads to further increase. Also known as a 'vicious circle'

power (1) The rate at which work is done; or the amount of energy converted per second. Power is measured in watts (W). 1 watt = 1 joule per second (Js^{-1}). (2) Electrical power is the rate at which electrical energy is converted into other forms. Often measured in kW; 1000 watts

precipitation reaction a chemical reaction in a solution, in which an insoluble solid is produced

prediction a statement about what you think is likely to happen

producer any green plant that produces food by photosynthesis

product a substance formed in a chemical reaction

proof evidence that is 100% convincing, or almost 100%

properties the characteristic features of a particular substance

protein giant molecule made of long chains of smaller molecules known as amino acids

protons positively charged particle in the nucleus of an atom

pumice a kind of rhyolitic rock that contains many gas bubbles, so has very low density

radiation electromagnetic energy

reactant substance that reacts with other substances and thus undergoes a chemical change

recessive in genetics, a gene that is not expressed if the dominant gene is also present

reduction any chemical reaction that results in oxygen being removed from a compound.

renewable energy energy from resources that will not run out

reproduction making more members of the same species

resistor an electrical component which converts electrical energy into heat energy. Circuitry symbol is a rectangle

respiration a chemical process of releasing energy from foods such as glucose

rhyolite igneous rock formed as a result of explosive eruptions

sacrificial (metal) a metal which is rapidly oxidised, and in so doing protects a less active metal with which it is in contact

salt compound with metal and non-metal elements combined

saprophyte (*sap-row-fite*) any living thing that lives and feeds on dead and decaying plant and animals. *Sapro* means rotten

sedimentation any process that builds up soft sediments in layers. Mostly this happens underwater, but some sediments (eg. sand-dunes) are wind-blown

series a method for connecting electrical components one after another. There is only one path for current to flow, and current goes equally through every component

SI Systeme Internationale, which uses units such as joules, watts, and kg; but not horsepower or miles or calories

significant figures the number of numerical figures given in an answer or measurement. In general, 3 significant figures is adequate

solenoid a coil of wire through which a current is passed. Solenoids are the basis of electromagnets

South Celestial Pole (SCP) an extension of the Earth's south pole into the night sky. The stars in the southern sky appear to rotate around the SCP as the Earth spins

specialised adapted or modified or equipped for a particular function or environment

species a group of similar organisms that are able to breed with each other and produce living, fertile young

spectrum (visible) band of colours seen when white light is separated by a prism

speed a quantity of motion calculated by distance/time. Measured in ms^{-1}

spores tiny reproductive cells from fungi and bacteria that float in the air

star a large ball of gas in space that produces energy by nuclear fusion

starch a plant polysaccharide made of many glucose units; not soluble

static electricity a build-up of electric charge on insulators

steel the name given to all alloys of iron

subduction the tectonic process in which one tectonic plate is forced underneath another

sucrose a disaccharide carbohydrate; commonly called table sugar

sugars small-molecule carbohydrates such as glucose and sucrose

supergiant a massive star

supernova a huge explosion at the end of the life of a massive star

surface tension the 'skin' on water because the polar water molecules are attracted to each other (cohesion)

sustainable (1) any human activity that meets the needs of the present, without reducing the ability of future generations to meet their own needs. (2) any process that can continue for a very long time without greatly altering the environment

syncline a downward fold in rock strata

tectonic Earth processes that involve the slow movement of huge plates of crust, resulting in earthquakes, volcanoes and mountain-building

theory (1) in science, music and maths: a big idea or 'model' that explains many facts, and makes good predictions. (2) an opinion; not a fact

ticker timer a device used to measure speeds. Connects to 50 Hz power supply, and makes dots every 0.02 seconds on a thin paper tape

transpiration loss of water vapour by plants, mostly through their leaves

trophic level an animal or plant's position in a food chain. *Trophic* means feeding. Green plants are at trophic level 1

vaccine a substance that is artificially placed in the body in order to stimulate resistance to disease. (Originally referred only to smallpox)

vacuum empty space with no substance in it

variables factors in a situation or experiment that may affect the results

velocity speed in a specified direction

virus tiny, infectious pathogenic agent that reproduces only in the cells of living things. Viruses have no cells, no ability to move, and use no energy

viscous sticky and 'gluey'; very slow flowing liquid

volcanic igneous rock that has formed on the surface from lava or ash or rhyolitic material

volts the unit in which electrical energy is measured. One V is the energy carried by one coulomb unit of charge

wairua (Maori) the spirit or soul of a person that exists beyond death

watt a unit of power. One watt is the transfer of one joule of energy per second. Symbol W

weight the force of gravity acting on an object. Usually measured in newtons (N). On Earth, the force of gravity is about 9.8 N per kg mass

work the amount of energy transferred from one form to another

Y-chromosome a chromosome that occurs only in males, and is passed only from father to son

yeasts microscopic fungi that ferment sugars to obtain their energy

Zealandia the continental-crust section of the Indo-Australian tectonic plate that has existed separately for about 80 million years

zygote a fertilised egg

ISBN: 9780170214667

Glossary

Periodic table of the elements

KEY

atomic number	26
element symbol	**Fe**
element name	iron

- metal
- metalloid
- non-metal

	1	2	3	4	5	6	7	8	9	10	11	12	13	14	15	16	17	18
1	1 **H** hydrogen																	2 **He** helium
2	3 **Li** lithium	4 **Be** beryllium											5 **B** boron	6 **C** carbon	7 **N** nitrogen	8 **O** oxygen	9 **F** fluorine	10 **Ne** neon
3	11 **Na** sodium	12 **Mg** magnesium											13 **Al** aluminium	14 **Si** silicon	15 **P** phosphorus	16 **S** sulfur	17 **Cl** chlorine	18 **Ar** argon
4	19 **K** potassium	20 **Ca** calcium	21 **Sc** scandium	22 **Ti** titanium	23 **V** vanadium	24 **Cr** chromium	25 **Mn** manganese	26 **Fe** iron	27 **Co** cobalt	28 **Ni** nickel	29 **Cu** copper	30 **Zn** zinc	31 **Ga** gallium	32 **Ge** germanium	33 **As** arsenic	34 **Se** selenium	35 **Br** bromine	36 **Kr** krypton
5	37 **Rb** rubidium	38 **Sr** strontium	39 **Y** yttrium	40 **Zr** zirconium	41 **Nb** niobium	42 **Mo** molybdenum	43 **Tc** technetium	44 **Ru** ruthenium	45 **Rh** rhodium	46 **Pd** palladium	47 **Ag** silver	48 **Cd** cadmium	49 **In** indium	50 **Sn** tin	51 **Sb** antimony	52 **Te** tellurium	53 **I** iodine	54 **Xe** xenon
6	55 **Cs** caesium	56 **Ba** barium	57 **La** lanthanum	72 **Hf** hafnium	73 **Ta** tantalum	74 **W** tungsten	75 **Re** rhenium	76 **Os** osmium	77 **Ir** iridium	78 **Pt** platinum	79 **Au** gold	80 **Hg** mercury	81 **Tl** thallium	82 **Pb** lead	83 **Bi** bismuth	84 **Po** polonium	85 **At** astatine	86 **Rn** radon
7	87 **Fr** francium	88 **Ra** radium	89 **Ac** actinium	104 **Rf** rutherfordium	105 **Db** dubnium	106 **Sg** seaborgium	107 **Bh** bohrium	108 **Hs** hassium	109 **Mt** meitnerium	110 **Ds** darmstadtium	111 **Rg** roentgenium							

6	58 **Ce** cerium	59 **Pr** praseodymium	60 **Nd** neodymium	61 **Pm** promethium	62 **Sm** samarium	63 **Eu** europium	64 **Gd** gadolinium	65 **Tb** terbium	66 **Dy** dysprosium	67 **Ho** holmium	68 **Er** erbium	69 **Tm** thulium	70 **Yb** ytterbium	71 **Lu** lutetium
7	90 **Th** thorium	91 **Pa** protactinium	92 **U** uranium	93 **Np** neptunium	94 **Pu** plutonium	95 **Am** americium	96 **Cm** curium	97 **Bk** berkelium	98 **Cf** californium	99 **Es** einsteinium	100 **Fm** fermium	101 **Md** mendelevium	102 **No** nobelium	103 **Lr** lawrencium

ISBN: 9780170214667